PLASTIC RECYCLING

PLASTIC RECYCLING

SATI MANRICH
AND
AMÉLIA S. F. SANTOS

Nova Science Publishers, Inc.
New York

LIBRARY OF CONGRESS CATALOGING-IN-PUBLICATION DATA

AVAILABLE UPON REQUEST

ISBN: 978-1-60456-969-8

Published by Nova Science Publishers, Inc, ✦New York

CONTENTS

PREFACE

Since the discovery of plastics several decades ago, the widespread consumption of plastic products and their subsequent inappropriate disposal and accumulation have recently generated new societal concerns of waste management due to their inherent slow degradability, high volume increase and low recycling rates, which are negative on the basis of self-sustainability. Regulations imposing waste reduction, reuse and recycling indices and responsibilities, as well as effective collecting system and the development of new, environmentally clean recycling technologies are some of the efforts to achieve the self-sustainability goals. The efficiency of the collection and sorting systems impacts directly on the amount of recycled plastics and on their cleanness and quality, therefore, enlarging their market potential. The development of new recycling technologies is diversified and can be classified into mechanical, chemical and energetic recycling. In mechanical recycling, successful technologies are achieved through the improvement of existing processes using additives, blends with other plastics and alternative processing routes in order to maintain the original properties of the virgin resin and even allowing them to return to the same application as originally intended. Chemical recycling processes to obtain intermediary products for new polymers become feasible due to the cost reduction of the raw materials involved. Lastly, despite the under use of the gross energy potential of the raw materials employed, energetic recycling plants are gaining a proportion of residues whose technological solutions for separation and/or reprocessing are deficient, but which, on the other hand, are voluminous, consequently solving the problem of both residue accumulation in densely populated regions and their respective insufficient energy supplies. In this chapter, the authors proposed to present an overview of the current state of this whole plastic recycling sector including their recent advances, and highlighting new markets and recent trends on recycling technologies around the world. However, mechanical recycling has been emphasized owing to the experimental and published work of Manrich's workgroup at the 3R Residues Recycling Center, which has concentrated on studying all the steps in the process of mechanical recycling.

Chapter 1

INTRODUCTION

For sustainable development and the limitation of environmental impacts to become a realistic goal, *reduction* of the currently growing consumption of non-renewable natural resources, *reuse* of products following consumption and appropriate *recycling* of discarded residues are of paramount importance. The effective practical application of the "3R" concept is especially important for the burning of carbon-releasing energy sources to be minimized. Some studies have indicated that, even if all emissions of CO_2 and other greenhouse gases were stopped immediately, the climate changes that have already occurred on the planet would remain for some decades. Since from the stand-point of thermodynamics and engineering practice, it is impossible to end such emissions altogether, efforts have to be made to reduce the risks to the environment, whenever the opportunity arises [1-3].

The reuse and various types of recycling of waste residues can lead to reductions in the use of non-renewable material and energy resources, with the energy savings generally ranked as follows: reuse > material recovery > energy recovery (energy from waste). Conversely, burying the residues in landfills, entailing as it does the total loss of material and energy, makes no such contribution [2, 4]. In the case of plastics, whose main current source of raw material is the petroleum, all recycling methods are technically viable and are briefly described next [5,6].

Mechanical recycling consists of the reprocessing of plastic residues into new products, different from or similar to the original products. The waste plastic used may come from the manufacturing process or from post-consumer products. This is the simplest way of recycling plastic waste, demanding the lowest initial investments.

Chemical or feedstock recycling consists of using heat or chemical treatment to break down plastic residues into their basic chemical components, the monomers or other products, which can then be recombined into polymers or used for other applications. Typical examples of tertiary recycling processes are hydrolysis and pyrolysis. Unfortunately, this kind of recycling demands huge investments and is therefore viable only for large-scale operations where the volumes processed are comparable to those in the petrochemical industry (thousands of tons annually).

Quaternary or energy recycling is the recovery of the energy bound in the plastic, by combustion, thus economizing on fossil fuels. However, the operation of such processes must guarantee that the emission of volatiles is controlled, to prevent the environment being contaminated by other paths [7, 8]. Recently, this type of process has been excluded from the normal concept of recycling and mentioned only as a form of energy recovery. In fact, this line of recycling is often regarded as a wasteful underutilization of the gross energy stored in the plastic.

Even though all the methods of material and energy recovery from plastic waste are technically feasible, in practice they encounter economic, legislative, market and other barriers. Published contributions in regulations, management and recycling of plastic waste using all forms of media are numerous around the world. Frequently, in books of edited contributions dedicated to plastic waste, each chapter deals separately with one of the topics covered. In this chapter, we present an overview of the current state of the whole plastics recycling sector, if somewhat sketchily in some areas, including a brief review of the recent research and development in the fields of mechanical, chemical / thermochemical and energy recycling of plastic waste. However, there is an emphasis on mechanical recycling, owing to the experimental and published work of Manrich's workgroup at the 3R Residues Recycling Center, which has concentrated on studying all the steps in the process of mechanical recycling.

Chapter 2

WASTE MANAGEMENT
OF PLASTICS RESIDUES

Since the emergence of plastics in the 1940s, which was impelled by their notable cost-benefit advantages over the traditional materials they replaced, the concepts of security, comfort and hygiene have been improved. In addition, their intrinsic characteristics of lightness, low processing temperatures, durability, low thermal and electric conductivity, transparency and flexibility, among others, had an immediate and growing impact on the correlated manufacturing sectors, reducing consumption of both energy and natural resources. Furthermore, the plastics industry was enormously successful in developing novel materials such as plastic wood, synthetic leather and paper.

On the other hand, the mounting volume of plastic residues, coupled with their extremely low biodegradability, generated a serious problem regarding the amount of space they took up. In the developed countries, the large urban centers have real difficulties in finding space for all the refuse, needing in some cases to transport solid waste over long distances to its final destination [9].

This problem, along with those arising from poor disposal methods and the associated environmental impact, the high added value of waste, the need to promote sustainable development of the production chain and to educate people to be more aware of the environment, has stimulated much research and practical activity in the fields of the recycling, degradability, reuse and reduced generation of plastic waste.

In view of the fact that plastic is said to compose between 5% and 10 % by weight of municipal solid waste (MSW) [10-12] and yet is the material of which the smallest fraction is recycled [13, 14], there is an ongoing discussion among government, society and the manufacturing sector on the apportionment of responsibility for the management of plastic residues in MSW [15]. The difficulty in recycling plastics arises largely from the big fraction of plastic products considered unsuitable for recycling from the outset and from the high operational cost of plastic´s collection systems.

Several mechanisms have been employed to increase the viability of the reversed logistics of returning the end-product packing material to the recycler: taxation of the manufacturing sector about government regulations [16] the establishment of taxes on non-recyclable packaging [17] the mandatory use of recycled materials in some sectors [9] incentives on the use of articles made entirely of such materials, with the removal of all licensing requirements on those who wish to produce them [18] the opening of new markets for recycled plastic, implementation of policies of exchanging post-used packages etc. for toys, spendable vouchers, or sports material in needy communities [19, 20] and lastly programs to inform, raise public awareness and provide opportunities for consumers to play their part [21].

In parallel, initiatives used to improve the recyclability of packaging by manufacturing plastics parts with fewer different resins and with easier separation of components that contain distinct resins, and reduce the use of multilayered material [13] adhesives, additives and labels on packs also play an important role [9].

Thus, in various parts of the world regulations have been adopted in order to achieve short and medium-term recycling targets [13]. In the European Community (EC), which became the European Union (EU) in 1992, the goals for rates of recovery and recycling in the packaging sector were set by the Directive 94/62/EC [22] to member countries, establishing June 2001 as the deadline by which these goals had to be reached: recycling of at least 25% and at most 40% by weight of all waste packaging and recovery of at least 50% and at most 65%. In addition, for each specific type of material, the fraction recycled should be at least 15%.

The available data confirm that these projected rates have been achieved in paper recycling in the EU [23]. Considering the plastic packaging sector, in Germany, the country that recycles the highest fraction of its waste in Europe, these targets have also been accomplished, or at least approximated, except in the case of composites. New EU targets for individual types of plastic waste have already been outlined for 2006, in which at least 20% of each type should be recycled [23]. These targets are currently being revised by the EU, but national governments, such as the UK, are also reconsidering their own packaging recycling and recovery targets for 2008 [21].

The particular concern over post-consumer plastic from the packaging sector can be explained by its short useful life, which reflects in its fraction 75% of all plastic waste [24]. In the EU, the recovery of plastic packaging residues was boosted mainly by improvements in mechanical recycling that resulted from better solid residue management practice [25]. The system of selective waste collection used in Germany, organized by Dual System Deutschland (DSD), is a worldwide reference [26].

In the EU countries in the decade from 1993 to 2003, generally speaking, the mechanically recycled fraction of all discarded plastic rose from 5.6% to 14.9%. Over the same period, largely because of the contribution made by energy recycling indexes, the fraction of plastic in landfill fell from 75.7% to 61% [14, 25, 27]. In Japan, the equivalent fraction for solid residues in sanitary tips is around 40% and much of this waste goes for energy recovery. The use of energy recycling in those countries is justified by the reduced combustion of fuels to produce energy, by the release of oil for the manufacture of virgin plastics and by the provision of an alternative source of energy that reduces the problem of energy shortage.

Similarly, in 2000, about 11% of all plastic produced in the USA was recycled. This represents a great advance, since a mere 1% of plastic residues were recycled in 1987 [9]. Nevertheless, this advance in recycling indexes has been achieved by formal recycling regulations [28]. Relating to Poly(ethylene terephthalate) (PET) recycling indexes, the USA is currently going through a period of stagnation in the recycling of this resin, according to the annual reports published by the American Plastics Council (APC), in spite of the very high recycling levels of PET achieved in 1995. At most, the amount of recycled PET is increasing in proportion to the growth in production of the resin [17, 29]. In 2001, the fraction recycled did not actually fall, but only because the slack in the home demand was absorbed by the export market [17, 30]. In 2004, the fraction of recycled PET was of the order of 21.6%, according to the National Association for PET Container Resources (NAPCOR) [31].

Turning to Brazil, in a national survey carried out by Plastivida, the plastics division of the Association of the Brazilian Chemical Industry (ABIQUIM), the proportion of plastic residues transformed by mechanical recycling is around 16.5%, higher than that in Europe [32]. Furthermore, Brazil is the third biggest market in the world for bottle-grade PET [18, 32] and the amount of this resin sent for recycling is of the order of 35% [20, 34, 35].

Given the precarious state of waste collection system in Brazil, such high rates of recycling are achieved only with the spontaneous involvement of low-income families whose earnings come largely from collecting plastic residues [20,

35]. This segment of the population currently represents about 500,000 informal workers [32].

Finally, another type of residue demands our attention: the great volume of rubber tires discarded annually in Brazil and accumulated annually around the world [36, 37]. Since the beginning of the nineties, many Federal and State Government ministries have been developing their own legal responsibility for this residue. In Brazil, resolution 258, passed by CONAMA (National Council for the Environment), obliges tire manufacturers and rebuilders (of retreads, remolds) to provide an environmentally correct destination for an amount of used tires proportional to their volume of production since 2002 and 2004, respectively [38].

Chapter 3

FIRST STEPS OF PLASTIC MATERIAL RECOVERY PROCESS: SORTING AND CLEANING

SORTING

One of the stages of plastic recycling that most threaten its feasibility as a productive operation is the sorting of plastic material from mixed waste and, especially, separation of the different types of plastic, which is hindered by the fact that quite different plastics may be used for the same end. In other words, a given product can be fabricated with very similar characteristics from distinct plastics and these act, in mixed residues, as impurities of each other after separation. This reduces the viability of the process and, in serious cases, can cause a whole production line to be shut down [39, 40].

Related problems that must also be taken into account are those of multicomponent items, good examples being car parts and plastic electric and electronic devices with embedded metal inserts, and multilayered products such as laminated, co-extruded or metalized flexible packaging [39-41].

The ideal practical solution to this problem would be to make suitable alterations in the plastic residues at source; that is, to redesign the original plastic product. At the design stage, then, priority should be given to reducing the number and variety of components in one product and the variety of materials employed as consumer goods with identical functions. Such ideas are exactly the opposite of current design trends, particularly in the packaging and disposable goods sectors. In most cases, the existence of multicomponent and multilayer residues is justified primarily on technical grounds, while the reasons for fabricating essentially the same product from diverse materials are based on economics and marketing [41]. It is thus hard to imagine the above-mentioned reversal in product design trends becoming a manufacturing priority on the grounds of purely environmental gains.

Nevertheless, on the positive side, all over the world we see a lot of effort being put into the research and development of appropriate technology that will minimize the problems caused by mixed plastic waste and varying materials. These studies concentrate on two fronts: the efficient separation of different plastics and other components, and optimization of the composition of compatible blends, or plastic composites of different materials, that combine, profitably, the distinct properties of the component polymers [39-46]. The first of these research areas will be discussed here and blends and composites in later sections.

The techniques used to separate mixed plastic residues can be classified in several ways, but here they will be grouped as *manual* or *automatic* and each will be approached in a specific way.

Manual Separation

The efficiency and productivity of this method depend entirely on the experience of the workers responsible for identifying and sorting the plastic residues. This is the method used in the vast majority of micro and mini-enterprises in developing countries, where manual labor is cheap, and in Materials Recovery Facilities (MRFs) worldwide, although it is still used even in some large organizations that recycle electro-electronic residues in developed countries [39, 42]. When training technical personnel in the manual separation of plastics, basic notions of how to distinguish between these materials have to be introduced. A systematic procedure for the identification of components of municipal plastic waste, especially the most prevalent plastics, which was proposed in an earlier publication of the Manrich's workgroup, has until now helped the Brazilian public to achieve this aim [39].

This method consists of three steps: in the first, the identification is made directly by codes; in the second, the identified product is correlated with the most likely material, and in the third, certain properties specific to each material are determined. The three steps are briefly described next.

- *Step 1*: Locate the identification code and note the number or abbreviation found in the recyclable plastic symbol: 1 = PET, 2 = HDPE, 3 = PVC, 4 = LDPE/LLDPE, 5 = PP, 6 = PS and 7 = others, where

HDPE, LDPE and LLDPE are high density, low-density and linear low density polyethylene, respectively, PVC is poly(vinyl chloride), PP is polypropylene, and PS is polystyrene. However, the residue from a product, part or component does not always display a code, which may be molded in relief or printed on the surface. If the code is missing or identified as 7, the procedure has to move on to the following steps.

- *Step 2*: Consult a table of data, such as table 1, which helps the user to identify the most likely material in a given product. It is found that, contrary to what would be environmentally correct, the number of most probable materials rises through the years, albeit rather slowly. Hence, from time to time, these tables need revising.

Table 1. Polymers used most frequently in fabrication of packaging material

Type of packaging	Typical use	Most probable material[*]
Bottles	Carbonated soft drinks	PET
	Cleaning materials and toiletries	HDPE, PP, PVC
	Cooking oil	PET, PVC
	Mineral water	PET, PP, PVC
	Vinegar	PP, PVC
	Yoghurt drinks	HDPE, HIPS, PP
Pots, containers and trays	Margarine	HIPS, PP
	Yoghurt	HDPE, HIPS, PP
	Sweets and chocolates	PET, PP, PS, PVC
	Disposable plastic cups	HIPS, PP
	Prepacked fruit & vegetables	PS, PVC
Lids	Soft drinks	PP
	Cleaning materials and toiletries	HDPE, PP
	Cooking oil	HDPE, PP
	Vinegar	LDPE
	Yoghurt	HIPS, PP
	Margarine	HIPS, PP
	Mineral water	LDPE, HDPE, PP
	Sweets and chocolates	HDPE, HIPS, PP, PS
Plastic bags	Supermarket bags	HDPE, PP
Films[†]	Fruit and vegetable bags	LDPE, LDPE/LLDPE, HDPE, PP
	Biscuits and snacks bags	LDPE, LDPE/LLDPE, PP

[*] While these materials are the most likely ones, the packaging can be made from others.

[†] Film is a term used for plastic sheets 254 μm or less thick, normally used in shopping bags.

- *Step 3*: Given the list of most likely plastics, determine some specific distinguishing properties that are simple to compare; these are indicated below, for the case of post-consumer plastic packaging. Here we will omit the techniques of differential scanning calorimetry and infrared spectroscopy. Depending on the plastics in question, identification may be achieved by testing only one of these properties, or a sequence of tests may be required. This sequence varies from case to case and an example will be given later, for the case of plastic bottles.
    - Transparency: transparent $\Rightarrow$ PET, PP, PVC, PS; translucent or opaque $\Rightarrow$ HDPE, PP, high-impact polystyrene (HIPS), LDPE/LLDPE, PET.
    - Whitening: exhibit whitening when folded $\Rightarrow$ PP, HIPS, PS, PVC.
    - Hinge: PP is the only plastic that withstands the repeated force used to open and close the pack with a one-part device, the hinge.
    - Density: the polyolefins, (HDPE, LDPE, LLDPE, PP) and expanded polystyrene (EPS) float on water, being less dense ($\rho < 1.0$ g.cm^{-3}).
    - Combustion: the flames and smoke given off by burning plastic are characteristic of each type. Table 2 describes these features of several plastics.
    - Solubility: ability to dissolve in various liquids or solvents is specific to each plastic. Table 3 shows some examples.
    - Halogen or Beilstein test: if a copper wire is heated to redness, rested on the plastic and then returned to the flame, and the flame turns green, the residue is a halogenated plastic, such as PVC, which contains chlorine.
    - Hardness and malleability: it is very hard to distinguish polyolefins from each other by means of simple tests alone. Experienced technicians differentiate the polyethylenes, LDPE and HDPE, from other plastics, including PP, as they are readily scratched with the fingernail, whereas the rest are too hard. Since LDPE is more malleable than HDPE and PP, it can be distinguished by bending or pressing the article.

Table 2. Behavior on combustion of the main polymers found in MSW

Material	pH of smoke	Odor of smoke	Color of flame	Ignition/ self-extinguishing
HDPE, LDPE, LLDPE	Neutral	Burnt candle	Yellow with blue base	Ignites
PS, EPS, HIPS, ABS	Neutral	Styrene smell / with much soot	Yellow with blue base	Ignites
PP	Neutral	Burnt candle	Yellow with blue base	Ignites
PVC	Acid	Acrid	Yellow with green base	Self-extinguishes
PMMA	Neutral	Methyl methacrylate (acrylic)	Yellow with blue base	Ignites
NYLON	Basic	Burnt hair	Blue with yellow tips	Ignites
PET	Neutral	Sweetish	Yellow	Ignites
PC	Neutral	Phenolic (carbolic)	Yellow	Ignites
PU	Neutral	Acrid, pungent, sour	Yellow with blue base	Ignites
Cellophane (regenerated cellulose)	Basic	Burnt paper or plant material	Greenish-yellow	Ignites

Table 3. Solubility of polymers in various solvents

Polymer	Soluble in	Insoluble in
HDPE	Decalin[*], tetralin[*], xylene[†]	Ethyl alcohol, chloroform, benzene, acetone
LDPE	Heptane[*], xylene[*], decalin[*], tetralin	Ethyl alcohol, chloroform, benzene or petroleum ether, acetone
LLDPE	Xylene[†], decalin[*], tetralin[*]	Ethyl alcohol, chloroform, benzene, acetone
PP	Xylene[†], decalin[*], tetralin[*]	Ethyl alcohol, chloroform, benzene, acetone
PET	o-chlorophenol[*], nitrobenzene[*]	Xylene, ethyl alcohol, chloroform, benzene, acetone, cyclohexanone, methyl ethyl ketone (MEK), dimethyl formamide (DMF), THF
EPS, PS	Chloroform, xylene, tetrahydrofuran (THF), ethyl ether	Alcohols
ABS, HIPS	Chloroform, xylene, THF, methylene chloride	Alcohols, benzene
PVC	Cyclohexanone, MEK, DMF, THF	Chloroform, alcohols, xylene
NYLON	Formic acid, phenol, trifluoroethanol, concentrated sulfuric acid	Alcohols, chloroform, xylene, acetone

Table 3. (Continued).

Polymer	Soluble in	Insoluble in
PC	Chloroform, cyclohexanone, DMF, cresol, methylene chloride	Alcohols, benzene, acetone
PMMA	Acetone, toluene, chloroform, MEK, THF	
PTFE	Insoluble (soluble only in fluorinated kerosene at 300°C)	

* Soluble at temperatures above 50°C.
† Soluble at temperatures above 100 °C.

Table 4. Practical test sequence suggested for plastic identification from a mixture of waste plastic bottles

Property	Transparent		Translucent		Opaque	
Probable material	PET, PP, PVC		HDPE, HIPS, PP		HDPE, HIPS, LDPE/LLDPE, PP, PET	
Property	Whiten when folded	Do not whiten	Whiten when folded	Do not whiten	Density < 1	Density > 1
Probable material	PP, PVC	PET, PVC	HIPS, PP	HDPE	HDPE,LDPE/ LLDPE, PP	HIPS, PET
Further tests needed to complete or confirm identification	Halogen test Density Comb. Solubility	Halogen test Comb. Solubility	Density Comb. Solubility	Density Comb.	Hardness Malleability	Comb. Solubility

The test sequence suggested for the specific case of bottles is shown in table 4. It must be said that manual separation of a mixture of PET and PVC bottles etc is quite easy, since carbonated soft drink bottles are all of PET and the rest can be differentiated by inspecting the base. If the bottle was injection molded as a preformed, then blow-molded with an injection-point at the middle of the base, it is made of PET; if it was extrusion parison blow-molded, with a weld-line across the base, it is PVC.

Automatic Separation

Automatic separation, like manual, is based on differentiating a given plastic in the waste from the rest by means of its physical or chemical characteristics. The main properties exploited are density, chemical structure, solubility, surface character and electrostatic and thermomechanical properties.

Density

In plastics recycling, apart from the traditional and modern hydrocyclones that separate materials by density differences as little as 0.01 g.cm^{-3}, there are simpler processes, with separating tanks containing aqueous solutions of various densities. The sketch in figure 1 shows how this might be done in the case of typical mixed municipal plastic waste.

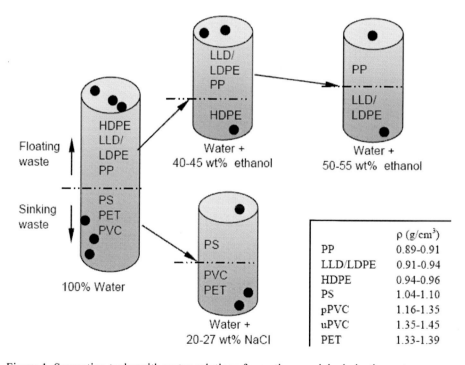

Figure 1. Separating tanks with water solutions for sorting municipal plastic waste.

Tsunekawa et al [42] have developed a process using density differences, based on experiments in the laboratory and in a pilot plant designed to separate plastics, recovered from scrapped copying machines, by a jigging method, which has culminated in a commercial plant, built recently to recycle office equipment, domestic appliances and automobile shredder residue.

Jiggers have been used in the technology of separation for a long time and are widespread mining sector. In this technique, water covering the material is pulsed at given amplitudes and frequencies, and the jolting of the solid particles results in their forming stratified layers arranged in order of density. In this plant, the TACUB Jig [42] was used and adapted to separate a ternary mixture of milled PS, acrylonitrile butadiene styrene (ABS) and PET, whose respective densities are 1.06, 1.18 and 1.71 $g.cm^{-3}$. Water density was raised by adding zinc chloride.

The pilot studies showed that, apart from the amplitude and frequency of the pulses in the water, other factors affected separation efficiency, including: height of the liquid (bed thickness), float level, water flow-rate and waveform. When performance was optimized by varying these conditions and analyzing the results, the plant separated 99.8% of the PS, 99.3% ABS and 98.6% PET for the ternary mixture.

Chemical Structure

Plastics can be differentiated and separated on the basis of atoms or bonds specific to their chemical structure, with the aid of devices based on infrared (IR) spectroscopy. Plastic residues, whole or ground into flakes, can be placed on a conveyor belt and scanned continuously by such a device, whose IR radiation source is adjusted to the absorption wavelengths of the plastic in question.

Any change in the absorption spectrum discloses the presence of an impurity at a specific point on the belt at that moment, thus defining the exact position and time, moments later, where a jet of compressed air should be activated. The foreign object is blown off the belt, to land on another, below the first, moving at right-angles to it and so effecting the separation [44]. By exchanging the IR source for one radiating in the ultraviolet-visible range (UV-VIS), the traditional technique used to separate glass by color can be employed. The advantage of equipment of this type is that plastic residues previously ground into flakes can be automatically separated, using different sources of radiation.

Solubility

The solubility profile of each plastic, already mentioned in connection with manual separation, is also the basis of automatic separation by selective dissolution. This can be used in two ways: either the plastic to be separated is extracted in a solvent that does not dissolve the others, or the residue is extracted batchwise with a common solvent that dissolves each plastic at a different temperature [45, 46].

Van Ness and Noster [46] presented the outline of a selective extraction process and the results of trials of the proposed method, performed by Linch Naumam with a mixture of known composition, containing virgin resins (LDPE, HDPE, PP, PS, PVC and PET), and a real mixture of post-consumer plastic residues. The efficiency of separation was found to be better than 99% in the case of the virgin resins, when tetrahydrofuran (THF) and xylene were used for selective dissolution at more than one temperature. However, it was not possible to separate PS from PVC at room temperature, or HDPE from PP at 160 °C, with THF.

In general, for the plastic mixtures discussed here, separation by density (see above) is more practical, as it is cheaper and environmentally friendly; however, it will not separate PVC from PET. When these are present, selective dissolution could be alternatively employed as a complementary final step, using xylene at 138 °C. In addition, solvent extraction is suitable for multicomponent or multilayer residues [41, 47].

Surface Properties

Most plastics exhibit low surface energy, making them hydrophobic. Despite this, the surface of plastics can be wetted by treating them with surfactant solutions, making them hydrophilic. The surfactants normally used for surface treatments consist of molecules with an affinity for organic polymers, but which also have hydrophilic groups. These surface modifiers are not equally efficient with all plastics, so that a given surfactant may be most effective for a specific plastic. These differences are exploited in the separation of residues by the method of froth flotation [48, 49].

The method is basically to put the treated plastic residues in a tank of water containing a frother and then introduce air-bubbles, which tend to stick to the hydrophobic particles, for which the surfactant was less effective. The clusters of bubbles and particles usually rise to the surface of the bath, in contrast to the hydrophilic particles, allowing the different materials to be separated.

Research on selective froth flotation focuses mainly on binary mixtures of PET and PVC [48, 49]. Drelich et al [48] applied the surface treatment in strongly alkaline solutions of caustic soda and used C_{9-11} ethoxylated alcohols as frothers in the flotation tank. The preliminary treatment made the PET hydrophilic, while hardly affecting the hydrophobicity of the PVC, and this was confirmed by measuring the advancing and receding water contact angles on each surface. The efficiency of separation attained was 95-100% in a variety of PET/PVC mixtures and flotation conditions.

In the work of Marques and Tenório [49], the surfactant used was calcium lignin sulfonate and the frother methyl isobutyl carbinol. Some of the experimental conditions were again varied, as were the sizes of PVC and PET particles. In the best conditions, 98.9% of the PVC and 99.3% of the PET were separated, at purities of 99.3% and 98.9% respectively.

Electrostatic Properties

The separation of several types of plastic packaging on the basis of differential triboelectrification was investigated by Hearn and Ballard [50]. The triboelectric effect is induced by rubbing two materials together so that, on separation, one retains a negative electrostatic charge and the other, a positive one. If two different plastics are rubbed against a third material chosen as a suitable reference, the charge generated on the latter should be positive with one plastic and negative with the other.

In the cited work, polyvinylidene fluoride (PVDF) and poly(butylene terephathalate) (PBT) were tested as triboelectric probes for the separation of mixed packaging residues of HDPE, PET, PP, PS and PVC. From the results, a separation line was proposed for domestic waste in which the PVDF and PBT probes were mounted in sequence along a conveyor belt. At the first probe, PP would be separated from the rest as the only component to be charged negatively by PVDF. At the second point, PET and PS would receive a negative charge from PBT and thus be separated from the remaining mixture of HDPE and PVC, which would be positively charged.

No method was suggested to separate PET from PS, but it was proposed that a visible light-based device could recognize and separate transparent objects like PVC from opaque ones such as HDPE.

Once again it is worth remembering that, to separate this kind of mixture, it would be preferable to try other techniques that are simpler, cheaper, more practical and efficient.

Thermo-Mechanical Properties

The mechanism by which a material suffers mechanical fracture, brittle or ductile, depends mainly on the temperature and rate of strain. If a mixture of two plastics is ground under specific conditions that cause each material to break in a distinctive way, the resulting particles will have different shapes and sizes that might be separable later by sieving or by means of air cyclones or hydrocyclones.

This way of separating binary mixtures of plastics can be applied when there is a temperature range where the two brittle-ductile transition temperatures are different from each other, in which the two plastics would break in different failure mode. In their study, Green *et al* concluded that the processing window of temperatures for selective impact grinding of a mixture of PET and PVC could be predicted from tensile stress tests and measurements of β-relaxation properties of each plastic [51].

Green *et al* [51] refer to two cases of processes, one in a patent and one used by a company, which produced particles of PVC smaller than PET after mixtures were crushed or subjected to impact grinding. In the case of cryogenic grinding, it was possible to recover PVC, 99% pure.

A more recent approach to separating mixtures of plastics by their thermal transitions, which took into account the effect of dirt present in the residues, was proposed by Saito and Satoh [52]. In this study, the differing thermal adhesion behavior of different plastics was analyzed by employing two unsteady heat conduction apparatus. One consisting of a plate to heat, press and then to pull the adhered plastic pellets, so as to determine the adhesion temperature and the effects of varying some parameters, and the other counter-rotating twin rollers were used to improve the efficiency of adhesion. Tests were carried out on a mixture of pellets of PET, LDPE, PS, HDPE and PP.

Generally, the thermal adhesion temperature of a polymeric substance is specific to each material and related to its glass transition or melting point. Saito and Satoh [52] showed that the size and shape distributions of the pellets affect adhesion, but within a characteristic temperature range for each material. Their proposed method succeeded well in separating a mixture of three plastics, PET, PS and PP.

While these authors and collaborators showed that dirt caused only a slight rise in the adhesion temperatures, it remains true that a real mix of residues is very different from pellets, especially in terms of the shape and size of the particles. The separation of wasted plastic flakes needs to happen before extrusion into pellets, which would remains as a great challenge to be solved by this methodology.

CLEANING

As previously mentioned, the collection of the residues from which used plastic is salvaged play an important role in raising the recycling indexes [31, 35]. On the other hand, that activity is also directly responsible for the extent of contamination of plastic residues, which in turn has a strong influence on the cleaning process and hence on the effluent discharged.

It is clear that plastics, during their use and disposal, come into contact with other compounds, and their composition may be changed by contaminants permeating through and impregnating the material [53]. Thus, before recycling it is necessary to determine the extent of contamination, the contaminating chemicals and the intended use of the end-product, in order to adapt the cleaning technique as appropriate.

Processes employing aqueous solutions are normally used to remove surface contaminants [54, 55]. However, these are not so effective in removing hydrophobic compounds or those that have migrated into the polymer matrix. These solutions generally contain caustic soda and surfactants, the proportions and flow-rates varying according to the type of material being recycled. Usually this is a continuous process in which the plastic, previously grinded into flakes, is vigorously shaken in the cleaning solution. The concentration of the alkali depends basically on the amount of glue and labels attached to the flakes [54]. The removal of adhesives is important, especially for polymers susceptible to degradation in acid conditions, because at reprocessing temperatures, adhesives decompose to acids compounds.

Apart from sodium hydroxide, alkaline reagents commonly used include calcium hydroxide (lime), potassium hydroxide, sodium silicate, etc. The preferred alkaline fluid for this purpose is normally a mixed solution of 33% sodium hydroxide and 15% potassium hydroxide (caustic potash), although the susceptibility of a given polymer to degradation in alkaline conditions has to be taken into account when choosing these reagents [56]. The concentration of caustic soda used is usually adjusted to provide the desired alkalinity [54].

The concentrations of surfactants used are normally between 75 and 200 ppm. These agents are important in many real situations where the surfaces are contaminated with soil or microorganism. To choose a particular cleaning agent, it is necessary to consider the soil type, oil residues and unusual paints likely to be encountered. Other factors, such as the cleaning equipment, pH, liquid or solid state, work safety, environmental laws and costs of disposal should also affect the choice of the correct cleaning agents [55].

The cleaning step, in most cases, takes between 5 and 20 minutes at temperatures up to 88 °C. Short times do not suffice to remove adhesives, while periods exceeding 20 minutes give a low return, while the use of baths at high temperatures facilitates the removal of glue [55].

In view of the nature of the cleaning process and its relevance to the quality of the recycled product, it is vital that recycling activities that involve washing procedures go hand in hand with an evaluation of the effluents generated and discharged during this stage, so as to avoid merely swapping pollutants. Furthermore, the need to comply with both the international code ISO 14000 and local obligations to preserve the quality of water (in Brazil written into the Federal Constitution) make it essential to analyze the effluent from mechanical recycling.

Therefore, on the basis of previous studies about which thermoplastics are present in greatest amounts in MSW [12, 25], the Residues Recycling Center, 3R-nrr, at Federal University of São Carlos (UFSCar) have conducted laboratory-scale tests to characterize the liquid effluent produced by the cleaning of the three plastics found in largest proportions in rigid plastic packaging residues, viz. PET and two polyolefins, high density polyethylene (HDPE) and polypropylene (PP).

The polluting load of these effluents demonstrated that they were comparable to that of domestic sewage with a medium concentration of pollutants. No significant differences in the effluent characteristics were found between the two types of plastic studied, except for those differences intrinsic to the cleaning processes (temperature, surfactant, caustic soda concentration, pH) or arising from contamination of the plastics (oils and fats, solids). Generally speaking, these effluents would need to be treated before being discharged into bodies of water or even into sewage systems. Therefore, any plans for implementation of plastic recycling units should include an appropriate destination for such effluents in accordance with their final composition [57].

On the other hand, selective collection of these wastes proved to generate a much lower polluting load, as reported by Heyde and Kremer [58], indicating the importance of this type of collection since, among other advantages, it reduces the pollution deriving from the recycling process. Yet another advantage of selective separation of waste is that it reduces the need for and concentration of the cleaning agents used during recycling, thereby reducing its environmental impact and cost.

Another aspect of the work performed in 3R-nrr is the use of the effluent characteristics to evaluate the process parameters in terms of cleanness efficiency. Good correlations were shown, since there is no addition of chemicals and all the performance was evaluated through only one batch [57, 59-61].

Chapter 4

MECHANICAL RECYCLING

As defined earlier, mechanical recycling is the process of converting discarded plastic into new products, principally by melting and molding. In this form of recycling, the macromolecular nature of the polymer is not destroyed, so that the degradation reactions that directly affect the physical and chemical properties of the polymer and, at times, its appearance, are minimized and controlled, irrespective of the processing method chosen. Nevertheless, chemical changes that occurred during the original processing and in-service use may have a negative effect on the quality of products reprocessed by mechanical recycling, in comparison with those manufactured from virgin resin.

Mechanical recycling involves several steps, which generally include the following: collection, separation, grinding for large or thick objects or agglutination for films and thin objects, cleaning of the plastic to eliminate organic matter, drying (particularly important for polymers that are hydrolyzed) and reprocessing (figure 2). Some of these steps are also important in chemical or energy recycling and may be taken out or added according to the needs of the material/object being recycled, the desired end-product and available conditions.

The importance of the collection, separation and cleaning steps has already been mentioned. The stage of breaking into flakes or binding together, which normally occurs between separation and cleaning of the plastics, is mainly done to reduce the overall volume taken up by the plastic residues and to promote the interaction of those residues with the cleaning solution [62]. However, the size of the flakes produced at this stage may interfere with the cleaning efficiency, as the mass-transfer rate between the liquid and solid phases depends on the velocity gradient of the liquid near the surface of the flake [63]. Also, during extrusion of the plastic, depending on the system used, the size of flakes in the feed can be a hindrance to the feeding process [64].

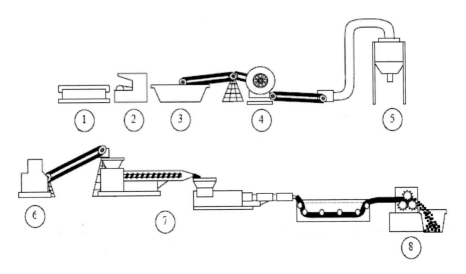

Figure 2. Complete recycling scheme (identification, separation and classification of different types of plastics (1); grinding (2); washing with or without addition of cleaning agents (3); drying (4); silos (5); agglutination (films and products with fine thickness) (6); extrusion (7); and granulation (8)).

After cleaning comes the drying step. This is mainly important in the subsequent reprocessing of molten polymers susceptible to hydrolysis and in the production of composites with inorganic fillers. The conditions chosen for drying depend, among other factors, on the type of humidity and the way it is bound to the material [65]. According to published work on this topic, there are three distinct kinds of moisture on recycled material: surface moisture, in the form of an external film, maintained by surface tension; free-moisture, which is inside nonhygroscopic materials, and bound moisture, either hygroscopic or dissolved, which exerts a lower vapor pressure than the pure liquid at the same temperature.

Materials classified as hygroscopic are those that contain water in a homogeneous solution in the solid phase, whose vapor pressure is below that of pure water. In these materials, water is normally extracted by diffusion. However, if the water content exceeds the maximum hygroscopic content, so that part of the water is free, then until the latter water is removed the material will behave as a nonhygroscopic material.

The drying conditions for hygroscopic solids are generally more aggressive than for others, since the lower the vapor pressure of the water, the higher its temperature of vaporization. By definition, the latter is the temperature at which the vapor pressure of the water is equal to the ambient pressure [66].

Consequently, drying processes for hygroscopic materials normally use high temperatures, with a supply of a suitable gas or a vacuum. The duration of this stage will depend on the temperature, the concentration gradient of water in the atmosphere, the relative velocity of the air, the residual humidity required and the pressure in the system. In vacuum drying, the concentration gradient and reduced vaporization point of the water, due to low pressure, are attractive features in view of the shorter time taken and the high quality of the end-product [66] if compared with other drying systems. Its chief benefit arises from the fast internal migration of the liquid resulting from the increased internal pressure gradient, which forces the water out to the edge of the material [67].

Finally, other factors affecting the drying time in hygroscopic materials are the size and shape of the particle and the degree of crystallinity. The bigger the particle, the larger is the path for water to migrate from the middle to the surface of particle. Conversely, the smaller the particle, the greater is the equilibrium water content, since the surface area available for absorption is larger [68].

Regarding crystallinity, the higher the proportion of crystalline phase in the polymer, the lower will be its final equilibrium water content. On the other hand, higher crystallinity also implies a lower diffusion coefficient of the water in the polymer matrix [68]. Owing to the first of these known correlations, commercial granules are normally sold in the crystalline state. Moreover, in polymers that crystallize relatively slowly, cold crystallization may occur alongside drying. In this case, especially if the recycled material contains flakes in the amorphous state, the drying is normally carried out with agitation or in a fluidized system, to prevent the particles sticking together because of the heat liberated during crystallization.

After the water content has been reduced, the material proceeds to the remelting / restabilizing step, which directly affects the eventual quality of the end-product. At this point, the polymer mixture is formulated in accordance with the target application. Besides stabilizers, various other agents may be added to the polymer: reinforcements, other kinds of polymer, coupling agents, flame retardants [69], foaming agents [70, 71] and so on.

Variations that have been employed to add more value to the recycled product and/or to widen its market niche include, for example, solid-state polymerization to recover the original properties of virgin PET, dissolution and reprecipitation to obtain high quality, very pure material [72, 73], electrospinning to make nanofibre from recycled expanded polystyrene (EPS) [74], civil engineering [75] and road construction [76] applications and synthetic paper.

While all these initiatives are highly relevant to the goal of raising the fraction of waste plastic that is recycled, only a few will be described in more detail in this section.

REMELTING-RESTABILIZATION

During reprocessing, it is necessary to limit the polymer degradation processes and thus guarantee its performance, to protect its added value. Degradation leads to molecular-weight variation, the inclusion of new groups in the polymer structure and undesirable changes in the appearance and properties of the material.

Hence, this mechanical recycling step depends crucially on studying the degradation and restabilization behavior of recycled plastics. Some researchers have reported that, during reprocessing, thermoplastics exhibit the same processes of degradation as the equivalent virgin plastics, but accelerated. This occurs because of structural elements introduced by previous thermomechanical degradation, thermooxidation and photooxidation, and sensitized oxygenated chromophore groups that are products of oxidation. These elements affect both the susceptibility of the polymer towards further degradation and its miscibility with others.

For this reason, polymer residues destined for mechanical recycling must not have been degraded to the extent that they have lost their original properties, nor can they contain high concentrations of sensitized oxygenated chromophore groups.

During reprocessing, the polymer cannot be stabilized to block the oxidizing action of these groups, nor can a degraded polymer be transformed into a high quality recycled product, since its original rheological and/or optical properties cannot be recovered all at once [77]. Restabilization only ensures that the reprocessing is performed without a significant advance in the state of degradation and that the degradation mechanism proceeds at a much slower rate than would otherwise be the case.

Another factor that contributes to the loss of stability of polymers while undergoing their first cycle of use is the consumption of protective additives in the materials during their production, service life and disposal. Thus, information on the resistance of recycled polymers to degradation, obtained by determining their susceptibility to oxidation and their residual stabilizer contents, would allow the stabilizing agents to be used more effectively and efficiently to block the advance of further degradation [78, 79].

Degradation Mechanisms

Degradation occurs by various types of reaction classified as follows: thermal degradation, mechanical degradation, chemical degradation, photodegradation, biodegradation, thermomechanical degradation, mechanochemical degradation and photo-biodegradation [78].

During reprocessing, the commonest types of degradation for most polymers are thermooxidation combined with mechanical degradation. In the special case of hydrolyzable polymers, hydrolysis is the predominant reaction and an extremely low residual water content is required, such as 20ppm (0.002%) for PET blow molding or injection molding.

Oxygen acts as a catalyst of several types of degradation reaction, according to published evidence from various studies [80, 81]. In fact, the chromophore groups responsible for yellowing in PET are formed by thermooxidation reactions in the polymer [82-85]. Research carried out by the authors and co-workers indicates that the discoloration of PET can be accelerated by the relative humidity of air [86].

Hence, the temperature, the shear rate, the atmosphere and water content have significant effects on recycled plastic processing, in accordance with the type of polymer and its main mechanism of degradation. However, irrespective of the predominant mechanism, the initiation reactions that lead to free radicals being formed take place typically *via* cleavage of either the main backbone or branches of the polymer. During the propagation of polymer degradation, the following may occur [87]:

- polymer free radicals recombine to form crosslinks or branches. On the whole, this propagation reaction does not predominate, since such free radicals have low mobility in the polymer matrix (slow physicochemical process of diffusion) and only rarely will they recombine. On the other hand, polymers with occasional unsaturated groups (such as vinyl, vinylidene) in their structure may favor this type of reaction under certain conditions of temperature and oxygen content in the atmosphere and in the presence of specific additives such as peroxides [88]. The predominance of this reaction is reflected in a drop in the fluidity or melt flow index (MFI) and intrinsic viscosity of a solution of the polymer (reduction in the free volume of the polymer chains in dilute solution) and increases in the polydispersity (PD) and viscosity of the melt.
- disproportionation of radicals to form linear products.

$$\underset{H}{\overset{CH_3}{\underset{|}{\overset{|}{\text{ww}C-CH_2\text{ww}}}}} \xrightarrow{\text{Inic}\,\text{Iniciator}} \underset{\bullet}{\overset{CH_3}{\overset{|}{\cdots\,C-CH_2\text{ww}}}} \xrightarrow{\Delta} \underset{CH_3}{\overset{CH_2}{\underset{|}{\overset{\|}{\text{ww}C}}}} + \overset{CH_3}{\underset{}{\overset{|}{\bullet CH}}} \tag{1}$$

- – liberation of monomers by depolymerization.
- – chain reactions *via* free radicals in the presence of oxygen. These reactions are widespread and give rise to peroxy radicals (RO$_2\bullet$) (Eq.2), which readily remove hydrogen from the polymer matrix by forming hydroperoxides (ROOH) and other radicals (Eq. 3) [78]. In turn, the hydroperoxides, initial products of oxidative degradation of polymers, are easily decomposed by heat or light (at near-UV wavelengths) to alkoxy (C-O$\bullet$) and hydroxyl (OH$\bullet$) radicals, since the O–O bond is weak.

$$\overset{|}{\underset{|}{\text{ww}C\bullet}} + O_2 \longrightarrow \overset{|}{\underset{|}{\text{ww}C-O_2^{\bullet}}} \tag{2}$$

$$\overset{|}{\underset{|}{\text{ww}C-O_2^{\bullet}}} + \overset{H}{\underset{}{\overset{|}{\text{ww}C\text{ww}}}} \xrightarrow{k_r} \overset{|}{\underset{|}{\text{ww}C-OOH}} + \text{ww}C\overset{\nearrow}{\underset{\big\}}{\bullet}} \tag{3}$$

As a result of the action of the disproportionation and chain reactions promoted by free radicals in the presence of oxygen, the proportion of high-molecular-weight polymer chains is observed to fall, resulting in reduced viscosities of both the melt and the polymer solution and a narrowing in the molecular-weight distribution (MWD) [88].

Polymers are thought to have a critical molecular weight (Mc), below which they become brittle. Among polyolefins, this is generally assumed to correspond to the point during degradation when the number average molecular weight ($\overline{M}n$) has fallen to around half of the original value [87]. Some authors claim it can be taken to be in the region of 10 kDa. For example, polycarbonate has an estimated Mc of 14 kDa and for polyamide 11, Mc = 15 kDa [89]. This critical value relates to the fact that scission of the polymer chain is confined to the amorphous regions and tie segments that connects the amorphous and crystalline regions. As consequences of chain scission in amorphous phase, the stress transmitted between adjacent crystalline zones decreases dramatically reducing the elongation at break. Additionally, the chain segments released from previously entangled regions are allowed to crystallize, causing a rise in the density (crystallinity) during oxidation. This last phenomenon, known as chemi-crystallization, may lead to the formation of surface fissures, which act as stress concentrators during elongation under tension or flexure [87].

Finally, processes that terminate degradation can occur either by the recombination (branching and/or crosslinking) or dissociation (chain scission) of free radicals, thus transformed into inert or stable products. If the dominant reaction is recombination, there will be a rise in the thermal distortion temperature and the creep and softening points, simultaneously with reductions in flexibility and in the elongation at break of the polymer [87]. Conversely, if dissociation predominates, there will be a fall in $\overline{M_n}$ and in the breaking tension. It follows that the type of reaction prevailing during degradation is what determines the eventual properties of the compound or polymer.

Concerning the kinetics of polymer oxidation, the rate of these reactions depends on a series of factors, apart from the structural properties of the polymer previously commented, including the duration of oxygen absorption, the oxygen content in the polymer, pressure, action of light, temperature and traces of transition metals. These reactions also characteristically exhibit induction periods, are self-catalyzing and can be inhibited or retarded with additives [87].

The reaction rate of oxidation (Eq.2) is controlled mainly by the direct attack of oxygen at the most vulnerable sites in the various chemical structures of the polymer, such as the immediate neighborhoods of unsaturated bonds or tertiary carbon atoms [78, 87]. The probability of these initial reactions to occur is higher when the polymer has more reactive hydrogen. A classic example of this is the lower oxidation stability of polypropylene (PP), compared to high-density polyethylene (HDPE), due to the presence in PP of a larger number of tertiary carbons. Thus, PP is rarely used without antioxidants.

The decomposition of hydroperoxides can be accelerated by traces of metals (Zn, Ti, Fe, Cr, Cd, V, Al and Cu from cable insulators, etc.), originating from impurities, pigments and residues of catalysts (polyolefins), acid residues (chlorine from $TiCl_3$, AlR_2Cl or M_gCl_2), fillers and antistatic agents. The catalytic power of the transition metals can be put in the following order: Ba < Mg < Al < Ti < Ni < V < Fe < Cr < Co [90].

These impurities can allow oxidative degradation of the polyolefins by chain scission to occur at lower temperatures (< 150°C), if the number of unsaturated groups is minimal (<<0.1% by weight) [88]. An example is PP formulations with stearates that contain transition metals, which suffer thermal oxidation at 125°C.

At low temperatures, the proportions of unsaturated bonds and tertiary carbons seem to be secondary factors compared to transition metals and acid residues. Nevertheless, if an olefin polymer has low crystallinity and a high content of unsaturated groups ($>> 0.2$ ωt%), such as diene terpolymers, e.g. ethylene-propylene-diene elastomers (EPDM), crosslinked networks can be formed at low temperatures ($<< 150°C$) [88].

Hydroperoxide decomposition reactions (Eq.4) induced by metals can be outlined as follows [78, 87]:

$$ROOH + M^{++} \rightarrow RO\bullet + M^{+++} + OH^-$$
$$ROOH + M^{++} \rightarrow ROO\bullet + H^+ + M^+ \text{ or}$$
$$2ROOH \xrightarrow{\text{traces of metals}} RO\bullet + ROO\bullet + H_2O \qquad (4)$$

The alkoxy radicals produced in these decompositions reactions, or by the recombination of *tert*-peroxy (PP), generally combine with H from the polymer matrix, forming alcohols, water and new free radicals (Eq.5), or suffer β-cleavage (Eq.6). Alkoxy radicals are less selective in their reaction targets than peroxy radicals, as they may oxidize tertiary, secondary or primary carbon atoms and unsaturated groups [87]. In practice, it has been found that β-cleavage of alkoxy radicals is one of the principal mechanisms of polymer degradation.

$$(5)$$

$$(6)$$

Another possible degradation mechanism is ionic scission, which happens in most polymers possessing heteroatoms in their chain. An example of this type of chain scission occurs in ester groups, which suffer a substantial amount of hydrolysis above the Tg, resulting in a rise in the number of carboxylic groups and small molecule fragments [91]. These fragments are usually formed by cleavage of the chain ends. In some cases, they are undesirable in the product that will be kept in the polyester package being reprocessed. One familiar contaminant produced in PET is acetaldehyde; its presence in aromatized and carbonated drinks is imperceptible, but it has a drastic effect on the organoleptic properties of mineral water [81].

Hydrolytic reactions are catalyzed by acid or alkali and their rate increases in proportion to the concentration of solvated H^+ and OH^- ions in dilute acid or alkaline solutions [92]. Acid-catalyzed hydrolysis involves the protonation of the oxygen atom of an ester group in the claim, following by reaction with water to form equivalent quantities of carboxylic and hydroxyl end-groups. It should be pointed out that this hydrolysis is self-propagating, since one of its products is an acid (Eq.7) [91]. In the case of alkaline hydrolysis (Eq. 8), the hydroxyl ion attacks the carbonyl carbon, again forming equivalent quantities of –COOH and –OH end groups.

$$—X-R— \xrightarrow{H^+} -\overset{+}{\underset{H}{X}}-R— \xrightarrow{H_2O} —RH + HO-Y— + H^+ \tag{7}$$

$$\text{\large\ww}X-C\overset{O}{\diagdown} \xrightarrow[B^-]{+ O_2} \text{\large\ww}X-\overset{O^-}{\underset{B}{C}}— \xrightarrow{H_2O} \text{\large\ww}X-H + \overset{O}{\underset{HO}{C}}— + B^- \tag{8}$$

Hence, contaminants of polyesters such as PVC and gummed labels, which release acids (hydrochloric and acetic, respectively) at reprocessing temperatures, serve as catalysts in the hydrolysis of ester groups. Similarly, if PET residues are not rinsed well after exposure to alkaline solutions during cleaning, the ester bonds are likely to be broken by base catalysis. In any case, irrespective of the prevailing type of catalysis, thermal degradation (Eq. 9) always takes place alongside hydrolytic degradation, during reprocessing. Consequently, the content of carboxylic end-groups rises in parallel with the falling molecular weight of the polyester [93].

$$\text{-RCOOCH}_2\text{CH}_2\text{OOCR-} \Rightarrow \text{-RCOOH} + \text{CH}_2\text{=CHOOCR-}$$

$$\text{CH}_2\text{=CHOOCR-} + \text{HOCH}_2\text{CH}_2\text{OOCR-} \Rightarrow \text{-RCOOCH}_2\text{CH}_2\text{OOCR-} + \text{CH}_3\text{CHO} \tag{9}$$

Since degradation reactions occur after diffusion of a reactive agent (O_2, H_2O, etc.), the rate of reaction in the solid state also depends on the shape of the residue, its morphology, crystallinity index, ambient relative humidity and temperature [91]. It follows that changes in the crystallinity of a polymer during its degradation are especially important with respect to its susceptibility to hydrolysis and chain scission [94, 95]. Indeed, it is possible to estimate a critical amount of crystallinity, above which the diffusion of oxygen and water drops, occasioning a fall in the rate of degradation. This behavior has been observed in many polymers [94, 96, 97], which exhibit the chemi-crystallization mechanism parallel to chain scissions.

To sum up, all the polymer degradation processes involve some oxidation reaction as a component, which varies little from one case to another and is limited by the rate of an initiation step. The degradation may occur through the cleavage, crosslinking or branching of chains. Usually, one of these predominates, depending on the composition of the material and oxidizing conditions (mainly temperature and oxygen) [88]. It should be stressed that, whatever the case, oxidation in solid polymers is restricted to small domains because of the relative immobility of the macro-radicals involved. Furthermore, the hydroperoxides, being unstable, tend to reinitiate new chains of oxidation within the domain that has already been oxidized [87]. Lastly, differences that exist in the behavior of different polymers are mainly due to the chemical bonds, functional groups and types of chain found in their structures, as well as the presence of impurities.

Restabilization Methods

The commonest means of restabilizing polymer residues is by adding stabilizing agents to the polymer during processing. This method is known as restabilization with additives, or external stabilization [78, 98].

In choosing an additive, a compromise has to be found between its effectiveness as a stabilizer and the restrictions of the application; for example, in food packaging, it must be non-toxic. Further requirements of an additive include: resistance to attack by oxygen, humidity and microorganisms; good mixing capacity, compatibility with other components, solubility; adequate knowledge of its effects on the physical properties of the polymer, including rheological, mechanical and electrical characteristics, etc., and on the appearance of the reprocessed plastic, before and after its exposure to heat and light; and, finally, a cost-benefit analysis [78].

The concentration of the stabilizer added to the polymer will depend on the polymer matrix, the type of stabilizer; the color, intended service life, shape and size of the product; the presence of other additives and quantity of residual additives from the original product(s); the extent of irreversible change already suffered by the polymer; the cost and other factors. To balance all these requirements and conditions, it is common to use more than one in the formulation, in a search for synergistic action [78]. Generally, the target is to achieve the highest ratio of performance to concentration. This approach is reflected in a yearly decline in the consumption of antioxidants for plastics processing during the period from 1977 to 1992 [99]. Notwithstanding this, the production of plastics continued to rise over this period, while the requirements concerning the performance and stability of the polymers became stricter.

Stabilizers are grouped into classes, according to their mechanisms of action: primary antioxidants; secondary antioxidants (hydroperoxide decomposers) and deactivators of metals; UV absorbers; sterically-hindered amine light stabilizers (HALS), among others [78].

Primary antioxidants intercept the free radicals (ROO•, RO•, etc.) formed at the start of the degradation process and thus delay their propagation. They may react with free radicals by an addition reaction (combination), electron transfer or, more often, hydrogen transfer. Thus they are classed as free-radical trappers, electron donors or hydrogen donors. Primary hydrogen-donating antioxidants, also known as thermal stabilizers, are the class most often found in practice. They neutralize peroxy free radicals, generating a by-product that is stabilized by resonance across four different forms that coexist in equilibrium. These resonant forms may react with other peroxy radicals, neutralizing them by forming inert products. The mechanism of these reactions can be seen in figure 3 [78, 98].

Chemical structures of primary antioxidants of the sterically-hindered phenol type are exhibited in table 5 [78, 98]. This type of additive is preferably used to stabilize the melt or, in some instances, as photo-antioxidants. However, they do not function so well in the latter role as their reaction kinetics are far superior in thermal processes [98]. It is important to note that the activity of these phenols falls dramatically above 220°C, the limit being around 230-270°C, above which they do not function. This is connected with changes in the mechanisms both of the oxidation of the polymer matrix and of the action of the phenols [99]. Other examples of primary antioxidants are some sulfur compounds, such as thiophenols, and secondary aromatic amines.

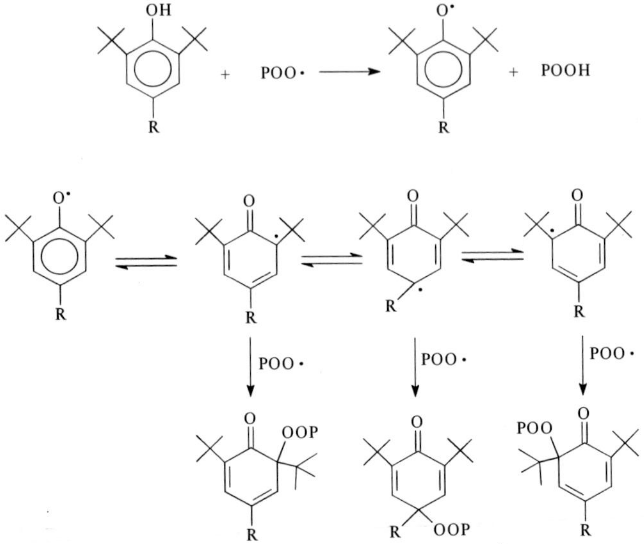

Figure 3. Outline of reaction mechanism of polymer stabilization by the sterically-hindered phenol class of hydrogen-donating antioxidants.

Table 5. Chemical structures of some primary antioxidants in the sterically-hindred phenol class

Chemical structure	Systematic and trade names
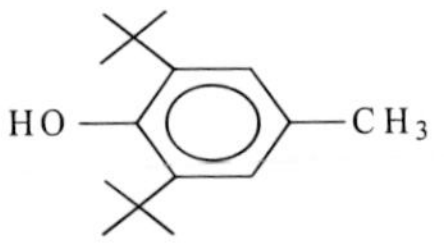	2,6-di-*tert*-butyl-*p*-cresol BHT (Uniroyal) and Ironol (Shell)
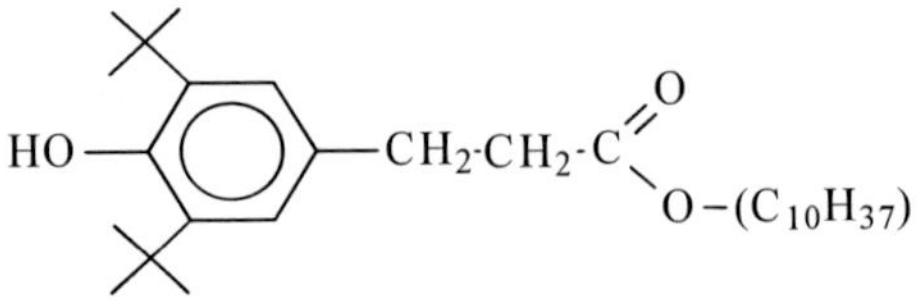	Octadecyl-3-(3.5-di-*tert*-butyl-4-hydroxyphenyl) propionate Irganox 1076 (Ciba Geigy)
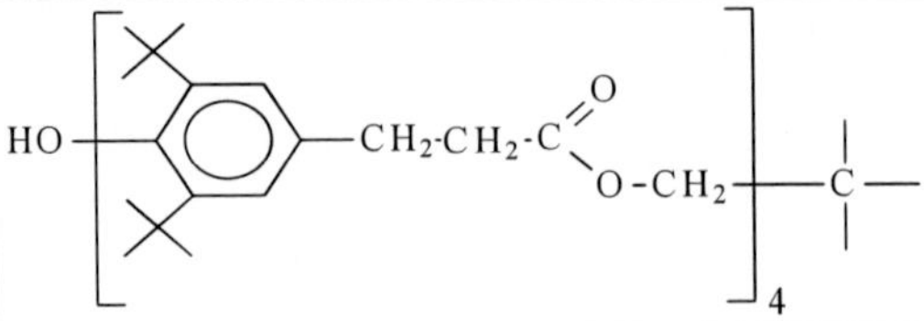	Pentaerythritol-tetrakis 3– (3.5-di-*tert*-butyl-4-hydroxyphenyl) propionate Irganox 1010 (Ciba Geigy)

Secondary antioxidants act on the initiation stage of thermooxidative processes, decomposing hydroperoxides into stable products. The simplified mechanism of their action is drawn in figure 4 [78].

1: organic phosphite [*] , $(ArO)_3P$:

$(ArO)_3P$ + ROOH → $(ArO)_3PO$ + ROH

    organic phosphite   organic phosphate

2: thioether [†], $(R_1)_2S$:

$(R1)_2S$ + ROOH → $(R1)_2SO$ + ROH

    thioether   sulfoxide (more efficient decomposer of hydroperoxide than

                original sulfide)

Figure 4. Simplified mechanism of action of hydroperoxide-decomposing stabilizers (secondary antioxidants).
*used with phenolic antioxidants, these are recommended for stabilization during processing.
† used with phenolic antioxidants, these are recommended for long-term thermal stabilization.

However, Neri et al [100] concluded that above 250°C the reactions shown here are very unlikely, given the low thermal stability of hydroperoxides, which normally decompose into the radicals RO• and •OH at such temperatures. Therefore, from their studies on PP, they proposed that the phosphites, during processing, countered the oxidation of the polymer by reacting preferentially with the oxygen, being converted into phosphates. Furthermore, they found that completely aliphatic phosphites are more effective stabilizers than aromatic phosphites, since the former consumed more oxygen. The proposed mechanism of the phosphite reaction with oxygen is thus:

$$(RO)_3-P \xrightarrow{\Delta E} (RO)_3-P*$$

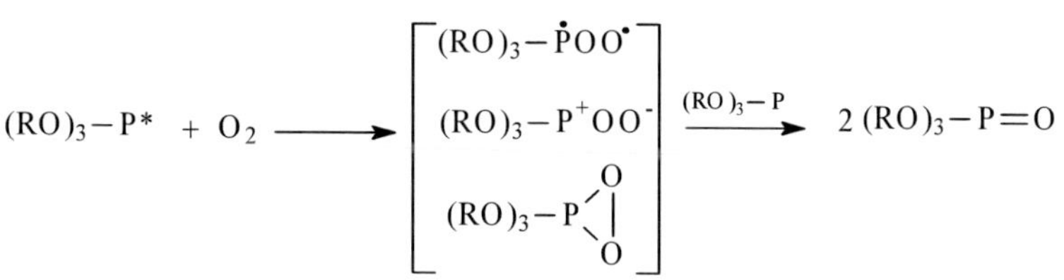

$$(RO)_3-P* + O_2 \longrightarrow \begin{bmatrix} (RO)_3-\dot{P}OO^{\bullet} \\ (RO)_3-P^{+}OO^{-} \\ (RO)_3-P\underset{O}{\overset{O}{\diagup}}| \end{bmatrix} \xrightarrow{(RO)_3-P} 2\,(RO)_3-P{=}O$$

$$(10)$$

The organic phosphorus compounds (phosphates and phosphonites) and thioethers (also called thioesters) are the antioxidants used most frequently. However, when using organophosphorus compounds, it should be borne in mind that they are readily hydrolyzed, generating phosphoric acid which may corrode the equipment.

Normally, combinations of primary and secondary antioxidants are used to obtain the benefit of the synergism between them in the thermooxidative stabilization of polymers [77, 78, 101]. This effect may be explained by a model in which the phenols convert peroxy radicals into hydroperoxides, slowing down the propagation reaction, while the secondary antioxidants decompose the hydroperoxides into stable alcohols. Even so, these antioxidants do not totally inhibit the oxidative degradation, but merely cause a substantial reduction in its rate of propagation [101].

A feature of stabilizing systems in the solid state is that the agents are normally segregated in the amorphous phase of the polymer matrix, as they have low solubility in the crystalline phase. This does not affect their effectiveness, however, owing to the similarly low solubility of oxygen in this phase.

In the vast amount of published research on polyolefin stabilization, there is a consensus that the best protection is provided by combinations of high-molecular-weight phenolic antioxidants with phosphites or phosphonites; in the case of PP, the phosphite : phenol ratio is usually 1:1 or 2:1, whereas for HDPE it can vary between 2:1 and 4:1 [77, 102]. Phosphite-phenol combinations not only maintain the molecular weight of the polymer, but also improve the shelf-life of the product, delaying natural and thermal aging by protecting and preserving other stabilizers [77, 79].

For the restabilization of recycled polyolefins, Popisil [79] recommends using about 0.05% more phosphite than would normally be used to stabilize the virgin polymer, in particular for PE. Similarly, Henninger et al. [77] suggest using a blend of stabilizers containing high phosphite: phenol ratios, to provide a safety margin for the protection of the residual additive systems found in most recycled material [73, 77, 79, 103, 104]. In this sense, when the present authors and co-workers conducted tests on post-consumer HDPE packaging, to establish the properties of the recycled plastic before and after reprocessing without adding any stabilizers, they found that the concentration of residual stabilizers in the plastic was sufficient to maintain its useful properties during the first reprocessing cycle [105].

Other synergistic combinations commonly found are: thioesters with phenols, which offer increased heat resistance of the polymer over prolonged periods; organophosphites with phenols, to increase stability during processing [106] and also to prevent discoloration, as organophosphites inhibit the formation of colored quinoid compounds from the phenols, when the latter donate their hydrogens to peroxy radicals; and aryl phosphite with a chain slipping agent, as a low-cost alternative system that proved to be as efficient as standard formulations in the reprocessing stage of PP recycling [107].

Metal deactivators are also important, as they neutralize the effects of metals that decompose hydroperoxides into active free radicals, promoting polymer degradation. Some compounds derived from hydrazine and hydrazone act as metal deactivators; for example, a *bis*-hydrazone that, by chelating metals, renders them unavailable to catalyze degradation processes.

According to Drake, cited by Malik *et al* [108], a mixed system of additives needs to be balanced, not only in terms of the chemistry of its components, but also in their molecular properties, such as the size, polarity, linearity and so on, of the constituent molecules. For Malik *et al* [108], in a "cocktail" of stabilizers, the amount and kind of mutual interactions among the constituents depend on the individual concentrations of the additives, the characteristics of the polymer matrix (including the mean molecular weight and its distribution), processing temperature, shearing rates, etc.

Specifically for PET, alongside published reports on the use of conventional antioxidants (phenol derivatives, aromatic amines) as stabilizers [81, 109], research has also focused on the use of chain extenders and solid-state polymerization to recover the original molecular weight of recycled PET. The adoption of restabilization procedures is of fundamental importance, since normally the degradation of recycled PET proceeds faster than that of virgin PET, even in trials in which the initial molecular weight of the virgin PET is lower than that of the recycled plastic [93]. Among the factors responsible for this difference are impurities in PET residues that catalyze hydrolysis and the shape and crystallinity index of flakes as opposed to pellets, which affect the amount of water absorbed initially: flakes absorb much more due to their larger surface area and lower crystallinity.

In reactive extrusion, the multifunctional chain extenders are designed to react with terminal carboxylic and/or alcohol groups in PET, counteracting the molecular weight losses caused by the reprocessing and concomitantly reducing the content of carboxylic acid groups (-COOH), when these are the site of attack of the additive [93, 109, 110].

Solid-state polymerization is performed at a high temperature under a flow of inert gas or reduced pressure, so that the polymer chain ends gain sufficient mobility to react with each other, the by-products being removed so as to displace the reaction equilibrium toward the product side.

In general, whatever procedure is used to overcome the degradation reactions, restabilization is an essential prerequisite to ensure the desired final performance of the recycled product, especially in the case of closed-loop recycling.

BLENDS

Research into and development of mixtures of different polymers has opened up a well-mapped path to the production of ´novel´ polymers and to the mechanical recycling of polymer residues, especially over the last two decades. This has occurred in spite of the unfavorable thermodynamics for miscibility: for most polymer pairs, the enthalpy of mixing is unfavorable and the entropy is negligible [111-115].

This being so, the blending of polymers inevitably leads to a heterogeneous system of multiphase morphology. When a blend consists of two components, A and B, and component B, say, is present in small amounts, the morphology is mainly of the droplet-matrix type, or discrete (dispersed) phase structure (DPS). Steadily increasing the proportion of B will lead to the percolation limit, characterized by a co-continuous or bi-continuous phase structure (BPS), in which each phase is connected continuously throughout the volume of the blend. Beyond this point, phase-inversion eventually occurs, with phase A in discrete drops in a continuous matrix of B [112, 114].

The physical and mechanical properties of a given polymer pair are strongly influenced by the multiphase morphology, which depends on the composition of the blend. Thus, it is possible to obtain a blend with desired properties by controlling its composition and hence its microscopic structure. For example, the BPS structure is particularly interesting when it is desired to maximize barrier properties, conductivity or impact strength [114].

Lyngaae-Jorgensen and Utracki [114] describe well-established correlations between microscopic structures of polymer mixtures and their bulk properties, such as Young´s modulus, electric conductivity, permeability, etc. According to Willense et al. [115], the elastic modulus, measured under tension, is mainly determined by the matrix phase when the blend has the droplet-matrix structure (DPS). Conversely, this modulus is determined by the dispersed phase when the DPS is of the fibrous type, especially in oriented specimens. The co-continuous BPS structures exhibit an intermediate behavior, with neither phase predominating, and the modulus is high and isotropic, owing to the interpenetrating phase structure.

However, the properties of a mixture of immiscible polymers, even though the composition may be fixed, can vary over time, since its structure may change as a result of not being stabilized soon after the blending stage, when the end-product was fabricated. Furthermore, the two or more phases formed by the mixture may show high values of interface tension and low adhesion between phases, impeding the transfer of stresses across the interface. This could result in a preferred path for fracture, along the interface, when the material is subjected to stress [116].

A third component may be added to reduce interface tension and make the phases more compatible: this would promote a structure with finer phases (smaller domains) and enable the development of technically interesting blends. The compatibilizing additive normally has three main functions: to reduce interphase surface tension by exerting an emulsifying effect; to improve interphase adhesion through its chemical affinity, enabling the transfer of stresses, and to stabilize the phase separation, preventing the dispersed phase from coalescing into larger domains [116, 117].

Several kinds of compatibilizing processes and reactions have been developed and applied to products made from virgin resin blends for decades, and more recently in the case of products from recycled plastics [116-127]. In this chapter we shall focus on blends of recycled material; published work in this field is found to concentrate on post-consumer packaging residues and scrapped electric-electronic and automobile parts [120-127].

Block and grafted copolymers are widely employed as compatibilizing agents; thus, the commonest of these used in blends of polyolefins with styrene-derived polymers is a styrene-elastomer copolymer [116-120]. Cherian and Lehman [120] report that the commercial use of such blends is growing in structural engineering, a prominent example being the recycled PS/HDPE blends incorporated into railroad ties and bridge beams.

Manrich's workgroup works with several blends, among them rHIPS/rHDPE, PP/PS, rPP/rHIPS, rPET/AES, PP/EVA and rPP/EVA, where r denotes recycled materials, aiming at recycling plastic residues into products of high added value. The accomplishment of such high quality recycled blends should be enabled by additive or synergistic combination of the properties of the blend components, as they are optimized for a specific application, in this case ecological synthetic paper for printing and writing. Some of these results are outlined in this chapter.

In polyolefin/styrene-derivative pairs, the compatibilizers used were block and grafted copolymers of styrene-butadiene, styrene-ethylene and styrene-propylene [118, 128-130]; in rPET/rPoliolefinas, ethylene-g-maleic anhydride copolymer [131]; in rPET/acrylonitrile ethylene-propylene-diene styrene copolymer (AES), MMA-glycidyl methacrylate (GMA) copolymer [132], while for PP/EVA, rPP/EVA and PP/rubber from scrap tires, no compatibilizer was added [133, 134]. In the case of rPET/AES and PP/rubber blends alone, the aim of blending was to toughen the thermoplastic.

We analyzed the effects of the composition and presence of MMA-GMA on the brittle-ductile transition temperature, impact strength and phase morphology of blends of PET recovered from beverage bottles with AES. The interphase modifier MMA-GMA is miscible with the SAN-rich AES phase and reacts with the polyester. It was shown to be indispensable as a reactive compatibilizer, causing a favorable modification of the morphology of recycled PET/AES blends and lowering the brittle-ductile transition point by up to 70°C, in the blend with 40% AES. At room temperature, the impact strength of this blend was increased 10-fold by this additive [132].

The remainder of this section is devoted to the investigation of blend intended for use mainly in ecological synthetic paper.

In the PP/PS blends, we used three interface tension reducers, PE-PS, PP-g-MAH and PP-g-PS, the last one synthesized for the purpose in this laboratory [118]. Two types of PP and two of PS were evaluated by micro-rheological tests in order to choose the best PP/PS pair for blending. Recycled plastics were not used at this stage, to limit the variability of the material.

The viscosity ratio of each of the four polymer pairs was determined. According to Navrátilová and Fortelný [135], the best polymer pair should be one whose viscosity ratio is lower than unity, as in this case the coalescence of the minor phase, which is less fluid, should be hindered more than if the viscosities were equal. In our study, however, all the ratios were fractional and an interesting observation was made when phase morphology was examined: the PP/PS blend whose ratio was nearest to 1.0 exhibited much finer dispersion of the PS phase than the others. Therefore, this blend was used for all further experiments [118]. Also, by determining the sizes of PS domains, it was concluded that PP-g-PS gave the best performance of the three compatibilizing additives.

In studies of PP/HIPS blends, made from recycled and virgin polymers, only one additive was tested: SEBS Kraton G was added at 5, 6 and 7 wt% to the compositions PP:HIPS of 2:1, 3:1 and 6:1 (w:w). The results of mechanical and morphological evaluations showed that the blends of recycled material had better properties than those made from virgin material and that the composition favored for the fabrication of synthetic paper was PP:HIPS = 2:1, plus 5% SEBS [128].

In the latest work on ecological synthetic paper, blends were prepared from rHIPS/rHDPE, rHIPS/rPP and PP/EVA [129, 130, 133]. In the blends based on rHIPS, comparative tests were performed on specimens containing the additives

SEBS Kraton G and multibloc SBS from BASF [129, 130]. In mixtures containing each of the additives, the glass transition (T_g) of the PS phase in HIPS and the melting peak temperature T_m of the PP phase were measured by differential scanning calorimetry (DSC). Whereas the T_m of PP did not change with the addition of either HIPS or the compatibilizer, T_g of PS was lowered.

In the case of mixtures of PP and EVA, preliminary tests on tubular films showed that these were compatible at low levels of EVA. For this reason, no interfacial agent was added. Indeed, a growing tendency in recent research is to try and use compatible polymer pairs, in concentrations at which a third component is unnecessary [40, 134, 136, 137]. One such study was developed in our laboratory with the aim of toughening virgin PP by adding powdered rubber (rubber shaving particles) from post-consumer scrap tires [134]. Neither additives nor compatibilizing treatments were necessary in the blends, and the results indicated a positive effect and close interactions between the rubber-plastic interphases when 5 to 10% of rubber particles were added to PP. Figure 5 shows the morphology micrograph for 15 wt% rubber addition.

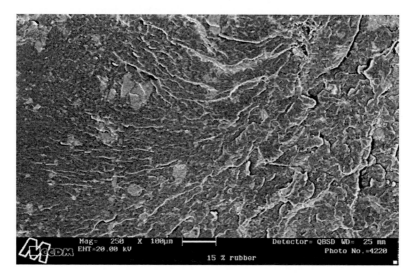

Figure 5. SEM micrograph of PP composite with 15% rubber shavings particles from scrap tires (250x).

Similar results were obtained recently by Scaffaro and co-workers [137]. They investigated the possibility of blending recycled PE (rPE) and post-consumer ground tire rubber (GTR) in order to obtain good mechanical performance blends without any additives. High temperature devulcanized GTR was also used in a comparison of two weight ratios rPE/GTR: 75/25 and 50/50. They found that good processability and mechanical performance was achieved at the lower amount of GTR in the blend. Moreover, the results suggested that high-temperature devulcanizing promoted reduced incompatibility by destroying the rubber cross-inking, though too high a devulcanizing temperature resulted in the viscosity, elastic modulus and tensile stress being increased by the carbonized rubber acting as a filler.

RECYCLING COMPOSITE PLASTICS

The proportion of composite materials that is recycled is one of the lowest, owing mainly to the lack of available technology that would allow these residues to be processed into viable products for new markets. Nevertheless, their economic impact is far from negligible, even if they represent a small fraction of all recycled plastics.

In theory, just as in the case of virgin resin, the addition of reinforcement to the recycled polymer matrix allows the composite to have a combination of properties that could not be attained by the individual components. Thus, while contributing to produce a composite with desired properties and therefore a product of higher aggregate value, the reinforcement is responsible for carrying most of the applied load, if the transference of mechanical load by the interface is efficient, resulting in the effective strengthening and stiffening of composite. On the other hand, the polymer matrix allows load to be transferred from fiber to fiber, prevents the catastrophic propagation of cracks between the fibers and protects the fibers against aggressive surrounding conditions such as humidity [138-140]. The effect of the reinforcement addition into recycled matrices on the mechanical properties of the resulting composite (Young's modulus, ultimate tensile strength, elongation at break etc.) is usually the same for conventional polymer composites.

The presence of fillers and/or reinforcements in the polymer matrix affects mechanical and rheological properties of the composite. The extent of this influence depends on many factors such as type and size of fiber/reinforcement, volume fraction, degree of dispersion and its interaction with polymer. The incorporation of fillers usually decreases tensile strength, while an opposite effect is observed by the addition of reinforcements. Elongation at break is decreased and viscosity increased by the addition of fillers and reinforcements due to the restriction of chains mobility [141-145]. These restrictions, on the other hand, may be beneficial to increase the Young's modulus (E), since for a same value of load applied to the composite, elongation is decreased, thus giving rise to higher modulus, i.e., composite stiffening and hardening.

Another important factor which is crucial for the composite to have a good performance is the polymer-filler/reinforcement interface. As previously mentioned, the load transference depends on this interaction. Therefore, when it is relatively weak, some approaches have been used to outcome this problem. The most common methods are the use of interfacial compatibilizers or coupling agents, which are substances used to promote interfacial adhesion and thus, improve the load transfer at the polymer-fiber/filler interface due to their compatibility with matrix. The mechanisms of compatibilization are based on modification of surface surface energy of reinforcements through chemical, mechanical or electrostatic adhesion with the matrix [146].

Reinforcing fibers in common use include both natural and synthetic fibers. The main advantage of using synthetic fiber is the possibility of producing fibers of unlimited length, while a natural fiber has a fixed range of finite lengths. Nonetheless, in applications calling for (that use) short fibers, many natural fibers compete successfully with synthetic varieties [146].

In the last few decades, natural fibers have been used increasingly, for reasons like low cost, the growing interest in exploiting materials from renewable instead of mineral sources, biodegradation of residues and non-toxicity. The natural fibers being widely incorporated into plastics recycled from MSW are that based on lignocellulose matrixes, in the form of wood fiber or wood flour. There have been various reports of the reactive extrusion of composites of plastic residues with this type of fiber, in which compatibilizers, flame retardants, foaming agents (when recycling the heavy fraction of MSW) and chemical treatment of the surface of the fibers are all employed [147-152]. Other natural fibers, albeit less exploited for plastics recycling, are those of jute, cotton, sisal and banana.

While there are innumerable reasons for using natural fibers, they also have disadvantages. For example, their hydrophilicity makes them less compatible with hydrophobic polymer matrices, and they show poor environmental and dimensional stability.

Organic textile fibers may also be used in polymer composites, despite their rather low stiffness, in order to reinforce matrices with even worse mechanical properties, such as rubber and thermoplastics. A good example is the fact that synthetic fibers made from recycled PET are capable of improving substantially the mechanical properties of PP [153].

When composites reinforced with glass fiber are recycled, they are first ground in a mill. Depending on the type of mill and extent of grinding, the recovered material may fall into two categories: (a) a fibrous fraction containing most of the reinforcement or (b) a fine powder containing most of the polymer matrix. It is important to separate these fractions, owing to the high added value of the used reinforcing fibers when they are mixed with a relatively small proportion of virgin glass fiber in new recycled composites [154]. In an alternative approach, chemical recycling is used to convert the polymer matrix into low-molecular-weight organic compounds and thus isolate and recover the valuable fiber fraction [155].

In spite of various other methods have been developed for the recycling of composite materials, the most widely used being to grind the composite scrap and incorporate it into a new composite structure, mainly thermoset molding structures. However, the inclusion of scrap during the actual fabrication of a recycled composite would lead to a very low-quality product unless the scrap content was kept low [156]. This limitation restricts the market of recycled composite that could be included in structural composites designed for the aerospace, naval and automobile industries.

One scheme that adds value to the recycling of thermoset materials, given that such residues cannot be melted during reprocessing and homogenized into the polymer matrix, consists of cryogenic grinding of the uncured flashes discarded during the respective molding process of the thermoset composite [157].

Finally, it should be noted that the extent of dispersion and homogeneous distribution of additives, reinforcement and fillers in the polymer matrix is greatly influenced by the type of process and the way it is carried out. Processes can be classified as continuous or batch, and also by the intensity of the shear stresses generated [64, 158].

In most batch processes, the degree of dispersion and distribution of system components is low, compared to continuous processes. This is due to the differences in stress and/or shearing rates that develop within the mixing chambers, generating large non-homogeneities in the dispersion and distribution of the particles. Hence, the end-product may have many distinct regions, in terms of the filler particle distribution.

Among continuous processes, extrusion is, without doubt, one of the commonest [158, 159], although a variety of alternatives does exist, such as injection, pultrusion, reactive injection of structural composites (SRIM), sheet molding compound (SMC), and others.

NEW MARKETS AND TECHNOLOGY

Recycled Plastics for Food Packs

The return of food packaging for reuse in the same cycle (closed-loop recycling) represents a victory for the plastics recycling industry, as it is responsible for a large fraction of all plastics consumed and also of the rigid plastics component in MSW, owing to the brief service life of food containers.

On the other hand, this type of recycling is a great challenge to the recyclers, as they are under obligation to ensure minimum potential risk to public health and not change the organoleptic properties of the packed food. Both before and after the plastic packs are discarded, there are possibilities that toxic chemicals, arising from the reuse of containers for other ends by consumers (storage of pesticides, herbicides, insecticides, household chemical products, car-maintenance products, solvents, disinfectants, products of microbial decay of food scraps, etc.), from adsorption and absorption of substances from the original contents and from contamination by contact with other residues, could migrate from the plastic to the food [27, 160, 161].

These possibilities are of real concern because of the non-inert nature of polymer matrix that allows sorption and diffusion of organic chemicals into plastics and due to the relative low temperatures of the recycling processes. Nevertheless, concerns related to the presence of microorganisms in the recycled plastic are neglected, as they adhere only superficially to the plastic and can easily be deactivated during recycling, by the pH of the cleaning solution and the polymer reprocessing temperature [162, 163].

In light of these considerations, the key criterion for the return of recycled plastic into direct contact with food regards to the assurance the decontamination of recycled plastic to a level that offers a negligible risk to public health and does not compromise the organoleptics properties of the packed food [164]. As ultra-conservative assessments, levels were set with reference to levels of carcinogenic substances at which short-term exposure offers a risk of contracting cancer below one in a million. Starting with these levels, a safety margin was added, so that the proposed maximum permitted level of contaminating chemicals migrating to the food was set at 2000 times lower than the cancer reference [161, 165]. Even after chronic exposure, this level includes a safety factor of 200 times. Based on the above discussion, the principle of "threshold of regulation" (T/R) was proposed, which established 0.5 ppb as the permitted limit for daily ingestion with negligible risk of a substance of unknown toxicity [165-167]. Hence, any contaminant in recycled plastic food packs which leads to daily ingestion of less than the T/R limit can be assumed inoffensive to human health.

Consequently, a food-packaging recycling process has to demonstrate its capacity to extract contaminants from the polymer matrix, treated here as food additives arising from indirect contact, reducing them to a residual concentration that does not expose the consumer to levels above the T/R limit. The evaluation of the process requires standard methods, to eliminate the great diversity of variables representative of the real contamination routes of these plastic containers.

These standard methods and its variations used to assess the suitability of a plastic recycling process for the production of food packs [27, 161-163, 168-176], consist, briefly, in contaminating flakes of packaging from post-industrial waste, free of any absorbed compound not inherent in its production, with model contaminants in a standardized and reproducible way and then submitting these contaminated flake through the recycling process in order to determine the level of residual contamination present in the plastic container produced by recycling. The recycled container or its food contents have to exhibit mean residual concentrations of the model contaminants equal to or below the T/R limit, for the suitability of the recycling process for production of recycled plastic packs for direct contact with food. Since evaluation of the residual contents of the model contaminants in the container takes no account of the barrier properties of the packaging, the process conditions determined by tests of migration of model contaminants to the food are generally more competitive. However, the complexity of the analysis and its costs are appreciably greater.

In the current view, conventional mechanical recycling is not considered suitable for the production of packaging to be used in direct contact with food, since the cleaning, drying and extrusion steps of these processes are inefficient at removing hydrophobic, polar and non-volatile contaminants, owing to the inbuilt limitations of each step [160, 161, 171, 173, 174, 177, 178].

In these circumstances, there now exist a variety of available techniques on the market, mainly concentrated in the fields of chemical and physical (monolayer and multilayer) recycling. Monolayer physical recycling methods, known as super-clean processes, are generally based on the steps of conventional mechanical recycling, to which further steps are added that employ heat, vacuum, inert atmosphere, solid-state polymerization [179-189], solvent extraction [190, 191], chemical surface treatments [192], vacuum degassing, supercritical fluid extraction (SFE) [193, 194] and steam distillation [195].These techniques have been tested experimentally and/or put into commercial practice, to enable recycled plastic to gain entry into the food packaging market.

Historically, the first super-clean process approved for use in direct contact with any class of food, under any conditions of use, appeared in 1994 for PET. The company that filed the application was Johnson Controls Inc. (Milwaukee – Manchester), the relevant division of which was later taken over by Schmalbach-Lubeca AG (Ratingen, Germany) [196-199], which nowadays belongs to AMCOR PET Global Packaging. Since that date, several other super-clean processes have been emerged; these are economically viable, but still run up against the limited quality and availability of the residues.

Recently, at the Residue Recycling Center at UFSCar, comparative tests of the rates of extraction of benzophenone from PET in dry air, vacuum or inert gas demonstrated that the diffusion of this contaminant from the plastic surface was fastest in dry air. On the basis of this result, a new super-clean process was proposed, which differed from others by using the conventional drying and crystallization recycling steps, at the upper temperature limit, to decontaminate the recycled plastic until reaching the residual contaminant level allowed for direct contact with food [200].

Besides the work reported in the patent, the authors and co-workers investigated the joint effect of using a liquid-phase extraction step followed by solid-state polymerization on the efficiency of contaminant removal. The liquid-phase extraction was carried out on d-limonene at atmospheric pressure and on water and acetone in subcritical conditions [201]. These results are shortly to be published.

In multilayer processes, two or three-layer packages are produced by co-extrusion of recycled and virgin polymer, the virgin layer of PET or PS, acts as a barrier to contaminants, since it is placed in direct contact with the food. In this arrangement, the recycled plastic can be made by conventional mechanical recycling, as it does not touch the food, but the disadvantage of this kind of process is that the fraction of recycled plastic in the pack is restricted.

Overall, PET is the polymer with the highest number of recycling techniques approved by the FDA, because the costs of the various super-clean processes developed are competitive with production of PET from virgin resin and the performance of the recycled PET, in some cases, approaches that of the virgin resin. Furthermore, the quantity of post-consumer PET collected is expected to increase, in tandem with the expansion of its market in the packaging sector [18], which will improve the viability of PET recycling processes, since the cost of installations rises more slowly than the production capacity [202].

On the other hand, food packaging applications that involve only limited contact at cool or refrigerated temperatures, a high ratio of the mass of wrapped food to the contact area, individually-packed food items or types of food that have to be washed before consumption, normally do not incur restrictions on the use of plastics obtained by conventional mechanical recycling processes. Examples of such applications are: recycled PS egg boxes, supermarket bags, recycled PE and PP crates and PET baskets for the transport of fresh fruit and vegetables [203].

Ecological Synthetic Paper

When articles of substantial added value are produced from discarded plastic, there is ample opportunity to market them as useful goods offering a viable return. This is particularly so when the sources of plastic are consumables (commodities); the recycling of these residues is much more efficient, as they are used and discarded in far greater volumes than durable goods.

In this light, we have put considerable effort into developing special films, designed to receive writing and print, from post-consumer recycled plastics. The films, known as *ecological synthetic paper,* would substitute synthetic paper made from virgin resins and even cellulose paper in some potential applications [204-206]. The motivation for this project is the possibility of finding one innovative and practical solution to several economic-environmental problems, three of which are especially urgent:

a) The need to create novel manufacturing techniques and new markets for goods produced from plastic residues, so as to minimize the damage to the environment caused by their disposal, given the clearly expected growth of plastics waste, considering the current growth trends in plastics consumption not only in developed countries but also, and particularly, in highly populated developing countries.

b) The need to meet the growing demand for cellulose paper; apart from the problem of insufficient supply for future demand, the current process of paper-making causes serious pollution, even though the raw material comes from a renewable resource. With regard to recycled post-consumer cellulose fiber, it is thought this could not be supplied in sufficient quantity at a viable cost-benefit ratio. As a measure of the scale of this problem in Brazil alone, the mounting investments in the paper and cellulose sector announced by the government for the years 2005 to 2014, intended to boost supply to the level of demand, are predicted to reach the vast sum of US$ 14.4 billion [207].

c) The crying need to reduce, perhaps even to stop, the cutting down of native forest, which occurs in part because of reforesting with monocultures of the exotic species preferred by cellulose manufacturers. This problem is particularly pressing in developing countries like Brazil, where, although the rate of increase in forest clearance in 2005 was lower than in 2004, we must not forget that 2004 witnessed the highest level of devastation of the Amazon rainforest in all history. Add to this the fact that for 2004, the projected timber requirements of the Brazilian cellulose industry were 135 million m^3 of planted forest, while the market forecast estimated availability at 125 million m^3, well short of the required volume [208].

Synthetic paper consists basically of laminated or co-extruded multilayer cavitied films with writing and printing properties similar to those of cellulose paper. It is practical, especially in industrial printing, when the supply of cellulose paper cannot fulfill demand, and when cellulose fiber hygroscopic properties are inconvenient. It may be used on several products: packaging including those for food, in personal documents manufacturing, maps, books, menus, billboards, envelopes, labels, instruction manuals, safe cards, pasteboards, paper currency, "busdoors" and "backbus" advertising panels etc [205, 206].

Practically all of the published information on synthetic paper is found in patents [209-217]. A basic feature of such documents is that the description of materials and methods is general and superficial. What can be gleaned is that multilayer films with the properties of cellulose paper can be made from thermoplastic composites with mineral fillers ($CaCO_3$, TiO_2 and SiO_2) and/or immiscible polymer blends, the polymer matrix consisting mainly of the polyolefins polypropylene (PP) or polyethylene (PE).

In a survey of recent patents, that of Squier *et al.* [217] refer to a thermoplastic film label that can be made by co-extrusion or lamination of uniaxially or biaxially oriented films, in five layers, comprising a central core (support), two adhesive and two external surface layers. In a general way, the patent include all of the principal materials, compositions and processes in frequent use. The printing processes covered include, among others, offset, silk screen, electrostatic and photographic methods, lithography, flexography, letterpress, thermal transfer and hot stamping. Inks mentioned are those employed in rotogravure, flexography and lithography, as well as in inkjet and hot melt printers, water and solvent based. It can be seen that this patent embraces practically every aspect of synthetic paper that has been protected in various patents around the world [217].

One very common feature in publications in general on synthetic paper is their reference to an important step in its fabrication, in which microcavities and microvoids are formed. Microvoid is a more general term, although it is often used as a synonym for microcavity. According to inventors, microcavities are small cavities that form along the interface between phases, owing to the delamination of layers by shearing during an extended orientation of multiphase film below the polymer melting-point. The dispersed phase of inorganic filler particles, immiscible polymer or even crystallites, may provide nuclei for the formation of new microcavities [210, 211, 218-220].

The mineral fillers, usually $CaCO_3$, are considered responsible for both the high mechanical rigidity of the films and the nucleation of microcavities. The latter, in turn, provide lower density, greater opacity and better printing qualities. The opacity is increased by microvoids as they provide a large number of points at which visible light are refracted in random directions. The dispersed filler particles distributed evenly over the surface, besides providing a key that anchors the ink, increase the roughness of the surface and so enlarge the surface area, leading to improved printability.

In the research done so far by Manrich's workgroup using mono-oriented and partially bioriented films, it has not been possible to prove these correlations between film composition/structure and printability. Several compositions have been studied, consisting mainly of post-consumer residues of PP and/or HDPE, to which another polymer component was added to improve printability, as well as inorganic fillers and white pigment [131, 205, 220-222]. As mentioned earlier in sub-section *Blends*, the plastics added to polyolefins to form blends were PS, HIPS, PET and ethylene vinyl acetate copolymer (EVA), along with compatibilizing agent in most cases.

Because of the limited availability worldwide of laboratory scale bioriented film-casting co-extrusion machines to introduce microcavities, our studies have been based so far on monolayer films fabricated by extruding flat mono-oriented films and blown tubular partially bioriented films. These films have not yet been submitted to tests that show whether co-laminating them into multilayer paper is a viable proposition [131,204-206, 220-222]. On the other hand, trials have been run very recently in a pilot plant on the co-extrusion of bioriented polypropylene (BOPP) in three layers, and the films obtained are currently being characterized in order to further being evaluated comparatively to commercial synthetic paper and cellulose paper.

Recently, a brief review of the work we had completed by then was published in the Proceedings of the international symposium REWAS'2004 [205]. We present next a summary of the work reviewed and more recent results from the ongoing research on ecological synthetic paper.

All recycled material was post-consumer; virgin PP, PS and EVA were resins supplied by *Polibrasil Resinas S.A.*, *BASF* and *Polietilenos União* resin companies, respectively. The plastics residues were all from discarded rigid packaging. The inorganic additives were as follows: $CaCO_3$ from *Imerys* and TiO_2 from *Polibrasil Resinas S.A.* Organic additives were: PP-g-PS (proprietary), PE-g-MAH from *Uniroyal*, SEBS and multibloc SBS from *Kraton Polymers*; thermal stabilizers / antioxidants *B 215FF* and *PS 802FL* (purchased from *Ciba*).

Conventional plastic waste recovery processes published elsewhere [128, 223, 224], including pre-grinding in knife mill, washing in caustic water solution, rinsing, drying, grinding into flakes or agglutinating into coarse powder, were used for the waste recycling. Treated and untreated $CaCO_3$ filler particles of different average sizes and size distributions were added, at 10-40 wt%, to PP, PP/EVA, PP/PS, rPP, rPP/rHDPE, rPET/rHDPE/rPPr, rPP/rHIPS and rHIPS/rHDPE (where r denotes recycled samples). The Drais high speed mixer and Baker & Perkins (b&p) or Werner Pfleiderer twin-screw extruders were used in this step, depending on each case. In the case of the blends, the polymers were premixed in single screw extruders with the addition of 0.2 wt% of thermal/oxidation stabilizers, prior to incorporation of the filler.

Flat dye casting extrusion (Gerst) and tubular film blowing extrusion (Ciola IF40) were the processes used to obtain the mono-oriented and partially bioriented polymer composite films, respectively. UV radiation treatment and a "solvent etching" were applied only to rPP/rHIPS [206] and rPET/rHDPE/rPP [131] film surfaces, respectively, in order to increase the concentration of polar chemical groups. Qualitative evaluations of visual appearance, writing quality and microscopic morphology, and several quantitative characterizations of physico-mechanical, optical, surface and printing properties were performed, in accordance with the following technical standards:

- Grammage, according to ASTM D646;
- Apparent density, according to ASTM D 792, when feasible;
- Tear propagation resistance and Tensile modulus and stress, following technical standard procedures ASTM D689 and ASTM D882, respectively;
- Optical properties - Opacity (ASTM D1003), Gloss (ASTM D2457), Whiteness Index ($L^*a^*b^*$)
- Friction coefficient, according to ASTM D4917;
- Water absorption, according to Cobb's Method – ASTM D779

- Printability characterization – Ink absorption and Ink adherence, through modified procedures according to ASTM D780 and ASTM D3359, respectively.
- Surface properties: Surface energy according to ASTM D724; Surface phase dispersion and morphology, characterized mainly through Scanning Electron Microscope.

As the surface free energy γ_s of a polymer is a property that will be referred to many times in this chapter, it will be convenient to explain it here. Its value is found indirectly from measurements of the contact angle (θ) or wettability angle of a drop of liquid on the surface of the polymer. This angle is measured between the solid surface and the tangent to the liquid-gas interface, at the point where the three phases meet. To calculate γ_S, equation 11 can be used [220, 225, 226], where γ_{SV}, γ_{SL} and γ_{LV} are solid-gas (polymer-air), solid-liquid (polymer-drop) and liquid-gas (drop-air) surface tensions, respectively.

According to the theory described in the references [220, 225, 226], the total surface free energy γ_S (γ_{SV}, eq. 11) is the sum of polar and dispersive components, i.e. $\gamma_S = \gamma_s^p + \gamma_s^d$, which can be estimated by measuring θ with two or more liquids, as long as the surface tensions and respective polar and dispersive components of the liquids are known. Two equations exist to calculate the surface free energy, referred as the harmonic mean and geometric mean formulae, but the former (eq. 12) is more often used for polymers.

$$\gamma_{SL} + \gamma_{LV} \cos\theta = \gamma_{SV} \tag{11}$$

$$\gamma_{SL} = \gamma_{LV} + \gamma_{SV} - 4\left(\frac{\gamma_s^d \gamma_L^d}{\gamma_S^d + \gamma_L^d} + \frac{\gamma_S^p \gamma_L^p}{\gamma_S^p + \gamma_L^p} \right) \tag{12}$$

where:

γ_S^d is the dispersive component of the solid surface free energy;
γ_s^p is the polar component of the solid surface free energy;
γ_L^d is the dispersive component of the liquid surface free energy;
γ_L^p is the polar component of the liquid surface free energy.

One of the main aims in much of the research carried out by Manrich's workgroup has been to reveal and comprehend possible correlations between the characteristics of the inorganic filler $CaCO_3$, or of polymers with high surface energy γ_S, and the mechanical and optical properties and printability of ecological synthetic paper films. We have analyzed the possible relation between γ_S and the printability of the specimens, as well as the possible effects of the type of process used and the process conditions on the formation of microvoids and the effects they have on the above-mentioned characteristics. The results of these analyses, presented elsewhere [205], do not allow any well-grounded conclusions to be drawn concerning these correlations.

Nevertheless, our investigation results did show that some of the variables studied had a more significant effect than others on properties of interest. Thus, to reduce light transmittance, opaque polymer has more effect than inorganic powder filler or white pigment; to enhance the specific offset ink printing properties, electric corona discharge (ECD) is more effective than thermochemical surface treatment and rHIPS-containing blends are preferable to PS-containing blends. Also, the simple presence of highly polar chemical groups on the film surface does not ensure ink absorption; rather, to achieve this, the surface polar groups and type of printing ink have to be compatible.

The possible correlation between the properties of surface energy γ_S, including its polar (γ_S^p) and dispersive (γ_S^d) components, and the absorption properties of printing inks on the ecological film paper have also been investigated. For synthetic paper composed of polymer blends and inorganic fillers, this is both a challenging and an interesting question to be elucidated. Thus, when ECD treatment is applied, provided the processing conditions and surrounding atmosphere are kept constant, the type and quantity of the polar groups formed by the electric discharge are critically dependent on the chemical structure of the polymer and probably that of the whole composite being treated [205, 225].

From the data cited [205] in table 6, it is seen that the replacement of PS by rHIPS in corona-treated films of PP composites increased the polar component γ_S^p and the offset ink absorption. This was attributed to a probable higher "susceptibility" to ECD shown by HIPS than by PS. Some data not published in the cited review, referring to tests on composite films of rPP/rHIPS blends, are also given in the table 6 [227].

The data in table 6 clearly indicate that rPET/rHDPE/rPP/$CaCO_3$ solvent-etched films exhibited lower offset ink absorption, even though their γ_S^p was higher than those of corona-treated films of rPP/$CaCO_3$ and PP/PS/$CaCO_3$ and similar to that of films of rPP/rHIPS/$CaCO_3$. This behavior was taken as evidence for our conclusion that high polarity is not the only and sufficient condition for the printability of the composite films. Qualitative writing tests, showing good absorption of different inks, e.g., pencil, organic solvent and ballpoint pen, corroborated these results for all samples.

Apart from the investigations referred to above [205], the influence of the concentration of $CaCO_3$ filler and ECD treatment intensity on the surface energy and its components was studied when the type of printing ink was changed from offset to serigraphy ink, for films of $rPP/rHIPS/CaCO_3$ composites. Possible correlations between the surface energy properties and the absorption and adherence of that kind of ink were analyzed [227].

A significant increase of γ_S and γ_S^p values with the surface ECD treatment and with its intensity was observed, as expected, and also a tendency for superficial energy γ_S to increase with the concentration of $CaCO_3$ filler. The dispersive component was only slightly affected by the intensity of the treatment, with a tendency to fall as the ECD level increased, in highly filled films. No correlation was observed between the absorption of serigraphy ink on the films and the values of γ_S and therefore between ink absorption and the concentration of filler. When filler concentration was maintained constant, the corona treatment affected significantly the absorption of ink on films of all compositions studied; however, the absorption increased as the intensity of the treatment increased only in composite films of relatively low $CaCO_3$ concentration, as can be verified in table 6.

Table 6. Surface energy (γ_S, γ_S^d, γ_S^p) and ink absorption (IA) of ecological synthetic paper

Films	γ_S (dyn/cm)			γ_S^d (dyn/cm)		γ_S^p (dyn/cm)		IA (g/m²)	
PP/CaCO₃ and rPP/CaCO₃						1.7 – 2.7		1.06 – 1.96	
PP/PS/CaCO₃						2.3 – 3.7		0.76 – 1.28	
rPP/rHIPS/CaCO₃						5.2 – 15.7		0.9 – 6.6	
rPET/rHDPE/ rPP//CaCO₃						8.88 – 14.78		0.03 – 0.30	
Cellulose Paper						21.7		14.6	
rPP/rHIPS/CaCO₃ ECD level (%V) Filler concentration	0	20	40	20	40	20	40	20	40
%CaCO₃ - 15	30.5	41.4	46.6	37.5	38.9	3.91	7.71	5.24	6.40
%CaCO₃ - 20	28.8	41.8	45.8	39.2	38.1	2.64	7.67	4.59	6.23
%CaCO₃ - 30	39.1	44.5	47.3	41.6	37.4	2.96	9.92	5.93	5.77

Similarly, by maintaining the filler concentration at a constant low level, it was found that both qualitative adherence and absorption test results were better for substrates with a higher polar component (γ_S^p) of surface energy and thus a more polar character, presumably due to stronger ink-substrate interactions [227]. On the other hand, for high concentrations of filler particles, it is possible that an effect of surface roughness prevails, so that equations 11 and 12 above are not applicable to these composites [228].

With regard to the study of microcavity formation and its effect on the properties of ecological synthetic paper, some recent results will be presented here.

Composite films based on rPP/CaCO$_3$ and rHIPS/rHDPE/CaCO$_3$ were characterized in terms of surface morphology, apparent density (ρ) according to pycnometry and theoretical density (ρ_t), whose mean value was calculated from the weight fraction and density of each component [221, 222]. The mean value of ρ was obtained from only two determinations, as there were no significant deviations. When calculating ρ_t, it was assumed that the degree of crystallinity of rPP and rHDPE did not vary significantly with composition or processing conditions, which had been shown to be true in earlier tests on composites based on PP and rPP/rHIPS. Surface micrographs were taken by scanning electron microscopy (SEM) with a field-emission gun, as a palliative means of observing whether microcavities were formed, given that in these 40 to 100 μm thick film samples it was not possible to measure them by quantitative techniques based on water or gas absorption or porosimetry.

In this work, two types of CaCO$_3$ were used, each one treated with the same interfacial agents, stearic acid and stearate, giving a total fatty-acid content of 1% according to the supplier information. The difference between them was granulometric, the particles of Carbital C110S ranging up to 10.0 μm, with a mean value of 2.0 μm, while those of Supermicro KRVA ranged up to 13.0 μm, with a mean value of 3.0-4.5 μm.

The results of these tests are displayed in table 7, where the following codes are used: P$_{cel}$ and P$_{sin}$ represent an A4 sheet of cellulose paper and a commercial synthetic paper; P and T signify flat casting mono-oriented and tubular partially bioriented films; rP and rH are for rPP and rHIPS/rHDPE–based composites; C and S for Carbital and Supermicro CaCO$_3$, filler, and the numbers 0, 10, 20, 30 and 40 show the filler loading in wt%. In all experiments, the blend rHIPS/rHDPE was mixed first, in the fixed ratio 80:20 (w:w), together with the compatibilizing agent multibloc SBS at 5 pphr (parts per 100 of resin).

**Table 7. Apparent density ρ and theoretical
density ρ_t of films of various composition**

Sample	ρ (g/cm^3)	ρ_t (g/cm^3)	$\rho - \rho_t$ (%)	Sample	ρ (g/cm^3)	ρ_t (g/cm^3)	$\rho - \rho_t$ (%)
PrPS0	0.90	0.90	0	TrPS0	0.90	0.90	0
PrPS10	0.95	1.05	−9.3	TrPS10	0.95	1.06	−10.3
PrPS20	1.05	1.22	−13.8	TrPS20	1.10	1.23	−10.9
PrPS30	1.11	1.36	−17.8	TrPS30	1.11	1.38	−19.6
PrPS40	1.22	1.49	−18.3	TrPS40	1.22	1.51	−18.9
TrPS30$_1$*	1.11	1.38	−19.6	TrPC30$_1$	1.11	1.36	−18.3
TrPS30$_2$†	1.07	1.38	−22.7	TrPC30$_2$	1.18	1.36	−13.3
				TrPC30$_3$‡	1.14	1.36	−16.0
PrHS10	1.13	1.26	−11.7	PrHC10	1,15	1.28	−11.5
PrHS20	1.21	1.42	−13.5	PrHC20	1.19	1.42	−19.4
PrHS30	1.29	1.56	−16.6	PrHC30	1.32	1.57	−19.2
Pcel	0.72	—	—	Psin	0.91	—	—

* Condition 1: film pulling rate 9.5 m/min
† Condition 2: film pulling rate 14.6 m/min
‡ Condition 3: film pulling rate 22.0 m/min.

The monoglyceride Dimodan, an antistatic agent used with polyolefins, was added at 0.2 wt% to rPP/CaCO$_3$ mixtures, while 0.2 wt% Dimodan and 1.0% titanium dioxide (TiO$_2$) were added to rHIPS/rHDPE/SBS/CaCO$_3$ mixtures. These additives were ignored in the calculation of ρ_t, owing to their relatively negligible proportions in the final composite.

The apparent density of the films rose continuously with the CaCO$_3$ loading, which was entirely as expected given that the inorganic filler (ρ = 2.6 g.cm^{-3}) is much denser than the plastic. However, it can be seen that, as the loading rises, the apparent density deviate further and further from the theoretical value, going from 10% below ρ_t at 0% CaCO$_3$ to around 18-19% below ρ_t in films with 30-40% loading. This implies that the CaCO$_3$ incorporated provoked an effect leading to a drop in density of the films, irrespective of which CaCO$_3$ (C or S) or which plastic matrix (rP or rH) was used.

This effect was observed even when the composite films were not submitted to biorientation, which, as pointed out earlier, should induce the formation of microcavities and thus would have furnished a ready explanation for the values lower than the theoretical densities. The films made from rPP with 30 wt% CaCO$_3$ were also drawn at various speeds, to test whether this parameter affected the quantity of microcavities and hence the density. It was predicted that the greater shear stresses produced at the polymer-filler particle interfaces by higher drawing speeds would lead to microvoids being formed in larger numbers and therefore to lower densities. Nevertheless, according to the data in table 7, no such effect occurred, as the film-drawing speed had no detectable influence on the apparent density. Furthermore, reducing the size of the $CaCO_3$ particles had no direct effect on ρ, irrespective of the polymer matrix.

Samples of the flat films PrPS30 and 40, PrHS20 and 30 and PrHC20 and 30 were subjected to extended biorientation in bench-scale tests. Only the microstructure formed during the orientation was analyzed, since the restricted size of the samples impeded the use of pycnometry or normalized grammage (grammage/thickness - $g/m^2/m$) techniques. This test procedure involved using a device, inserted in a muffle, capable of stretching the film uniaxially at a controlled temperature, as shown in figure 6. The stretching was applied transversely across the films up to four times, at 135 ^{0}C for films based on rPP and 110 ^{0}C for rHIPS/rHDPE films [221, 222].

Figure 6. Extended film biorientation device: sample holder (left) and the heating muffle (right).

Irrespective of the polymer matrix, composition, filler type, kind of extrusion (flat or tubular film) or the speed at which film was drawn, no microcavities were seen in any specimen, at least near the surface. On the other hand, microvoids due to shear-stresses at the polymer-filler particle interface were observed at the surface of every film submitted to extended biorientation. Figures 7a, b and c show this behaviour clearly.

The composite films of rPP/$CaCO_3$ were evaluated for possible use as a supporting core layer in multilayer films, so that there was no need to determine their optical and printing properties. Conversely, the PP/EVA/$CaCO_3$ and rHIPS/rHDPE/$CaCO_3$ films were studied as potential surface layers, whose printability and appearance would clearly be of supreme importance.

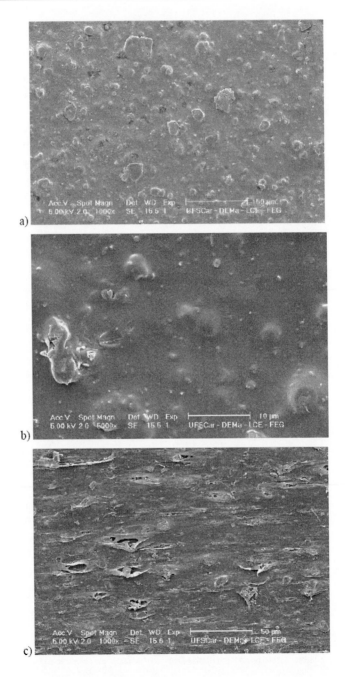

Figure 7. Surface SEM micrographs of PP/CaCO 60/40 films: (a) flat casting (1000x); (b) tubular partially bioriented (5000x); and (c) extended bioriented (1000x).

The flat mono-oriented films of rHIPS/rHDPE/CaCO$_3$ produced in the laboratory were tested in several ways. For example, it was found that both the type and concentration of CaCO$_3$ affected the adherence of offset printing ink. The ink was applied at Gráfica Suprema, São Carlos, Brazil, by inserting samples of ecological synthetic paper into a routine offset print run. The adherence of the ink was tested by fixing a length of adhesive tape (1.5 cm wide, Adere, Brazil) across the printed film, allowing a 560 g roller to run three times over the tape and then pulling the tape off the film rapidly.

The adherence of the ink to the synthetic paper was classified qualitatively as bad, good or excellent, according to the amount of ink removed by the tape. The results are shown in table 8. After this test, the specimens were photographed as a visual record of the quality of adherence. The worst and best cases are shown in figure 8.

Table 8. Adherence of offset ink to rHIPS/rHDPE/CaCO$_3$ films

Film	Without surface treatment	With surface treatment	Film	Without surface treatment	With surface treatment
SM10	Excellent	Excellent	C10	Bad	Bad
SM20	Excellent	Excellent	C20	Bad	Good
SM30	Excellent	Excellent	C30	Bad	Excellent
Pcel	Excellent	-	Psin	Excellent	-

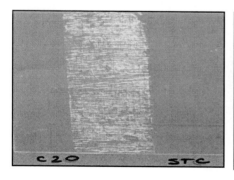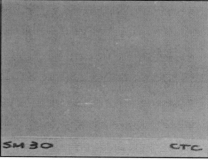

Figure 8. Visual illustrations of the worst and best cases of *Offset* ink adherence on rHIPS/rHDPE/CaCO$_3$ films: (a) C20 without ECD, and (b) SM30 with ECD treatment.

In the case of PP/EVA/CaCO$_3$ composites, the computer program MINITAB was employed to plan a factorial experiment, to analyze, with the help of a response surface, the influence of the inorganic filler and antistatic agent on the whiteness, coefficient of friction and quality of offset printing. These experiments were used to optimize the proportions of these components in the film for its intended use as a surface layer in ecological synthetic paper, and we hope to publish these results in the near future. Furthermore, optimized compositions have already been tested in a pilot plant, biorienting the PP (BOPP) by coextrusion in three layers. Only virgin raw materials were used in this test.

Results from the various tests used to characterize these three-layer films are still under analysis. While the measurements of gloss showed values within the accepted reference values, the surface appeared rather shiny, typical of plastic films. I t is planned in the near future, another pilot plant trial will be made, with the same compositions as before, except that post-consumer plastics will be used in all three layers, in the hope of reproducing in the pilot plant the normal appearance of synthetic paper, without a plastic shine, obtained in lab tests by extrusion of tubular film composed of recycled residues.

Chapter 5

CHEMICAL RECYCLING

Chemical recycling consists in using depolymerization and decomposition reactions to convert polymers into low molecular-weight products. According to the type of reaction, a variety of chemical products are generated that can be used as raw materials in several sectors. Chemical recycling is generally divided into two types: thermolysis and solvolysis [229, 230].

Thermolysis, also termed thermochemical recycling, works with high temperatures, from 350 °C to 1000 °C. It is based on three main processes: pyrolysis, gasification and hydrogenation. Solvolysis, which is often referred to simply as chemical recycling, involves the use of solvents and much more moderate temperatures than those used in thermolysis. Several kinds of solvolytic reaction are found in practice: hydrolysis, glycolysis, methanolysis, aminolysis, alcoholysis and acidolysis.

Solvolysis is chiefly used to recycle poly (ethylene terephthalate) (PET) [229-231], poly (butylene terephthalate) (PBT) [232], nylon [233] and polyurethane (PU) [234, 235]. Up to now, the industrial processes most frequently used to recover PET have been glycolysis and methanolysis. Nonetheless, hydrolysis has been attracting a lot of attention in recent years, as it can be classified as a green technology due to its simplicity, low energy consumption and low environmental impact. Furthermore, the fact that terephthalic acid (TPA) is currently taking over from the traditional monomer dimethyl terephthalate (DMT) in the synthesis of PET means that hydrolysis yields a readily usable raw material, even if the rather high cost of purifying TPA and ethylene glycol (EG) is a disadvantage of this process [230, 236].

Hydrolytic depolymerization of PET can be done in alkaline, acid or neutral conditions. Acid hydrolysis causes considerable problems, both on the economic and environmental fronts, due to the concentrated acids used. On the other hand, neutral hydrolysis, which is performed with hot water or steam at high pressure, is the most environment-friendly method, but it produces the TPA of lowest purity. Alkaline hydrolysis is generally carried out in aqueous solutions of sodium or potassium hidroxides, but other solvents may be used, especially when PET is salvaged from highly contaminated residues such as metalized and X-ray films and magnetic recording tape.

Karayannidis and co-workers [231] have developed a method of recycling spent PET by alkaline hydrolysis, which can be applied on the industrial scale to multilayer or contaminated residues, using 0.1 M KOH in methyl cellosolve. Assuming that the co-monomer isophthalic acid does not constitute an impurity, the TPA is found to be 99.6% pure. The TPA produced was tested by forming a co-polyester with EG, the result being a pure white polymer of intrinsic viscosity $\eta = 0.54$ dL/g.

When discussing recycling by solvolysis, an important aspect is the research effort being spent on novel media such as supercritical fluids. In this field, Goto and co-workers [237] developed a method of recycling PET waste by depolymerizing it in supercritical methanol at a pressure of 20 MPa and temperature of 300 °C, in an atmosphere of N_2. A continuous kinetics model was developed, tested on the experimental data and proved to simulate changes in the molecular-weight distribution (MWD) and monomer concentrations over time. This model included decomposition reactions such as random and specific scissions and secondary reactions for monomer components. The largest yields of DMT and EG were obtained at 7,200 s, 80 mol %, and at 3,600 s, 60 mol%, respectively.

The first of the thermochemical recycling methods, pyrolysis is carried out in an inert atmosphere, at 350-600 °C, and generates chemical feedstock with a composition similar to crude oil or even to naphtha, depending on the polymer residue used. Generally the liquid fractions predominate, the type of reactor, dwell time and process temperature being crucial to the product selectivity [229, 230, 238, 239].

Gasification is done at higher temperatures than pyrolysis and in an oxidizing atmosphere, the main aim being to produce syngas, a mixture of hydrogen and carbon monoxide used to synthesize basic organic chemicals. The material is commonly raised to 1000 °C and gasifying reagents such as oxygen, carbon dioxide, air or steam are introduced in separate or mixed flows. This technique is normally, but not exclusively, used on the residue of plastics already decomposed by pyrolysis [229, 230, 240].

Hydrogenation is effectively the cracking of polymers with hydrogen gas, at high temperatures, to convert them to liquid fuels. In this process, previously depolymerized plastic residues are submitted to cracking at around 480 °C and a pressure of 200 bar, generating various products, including gasoline and diesel oil, whose economic value is considerably higher those obtained by pyrolysis and gasification [229, 241, 242].

Judging by the published reports consulted, there was a period of intense research and development in the field of thermochemical recycling of plastic by the techniques outlined above between 1995 and 2000, focusing on plastic waste in general [238-243]. After that period, however, such publications suddenly thinned out as researchers began to concentrate on one kind of plastic, a current industrial problem or some specific question not resolved in earlier studies [244-246].

For example, the thermolysis of polyesters like PET and PBT had always given problems due to sublimation of the acids produced by the decomposition, which corroded the reactor and could also neutralize the lime (CaO) used as a chlorine filter, even blocking the pores in the filter bed. In addition, the sublimated TPA itself, by condensing in cooler regions, caused blockages in the equipment and generated an undesirable fraction of solids [230, 246].

Yoshioka and co-workers [246] considered an interesting solution, in which the pyrolysis was performed in a fixed-bed reactor in the presence of calcium hydroxide ($Ca[OH]_2$), and carried out experiments on PBT and poly (ethylene naphthalate) (PEN), similar to previous tests with PET, but now using helium gas as the inert atmosphere. A benzene yield as high as 85% had been achieved by decomposing PET at 700 °C in a steam atmosphere. In the case of PBT and PEN, respectively, a benzene yield of 67% at 700 °C and a $Ca(OH)_2$/PBT ratio of 10, and a naphthalene yield of 80% at 600 °C and a $Ca(OH)_2$/PEN ratio of 5, were obtained along with reduced carbonaceous residues.

These authors attributed the difference in benzene yield between PET and PBT to hydrolytic degradation of PET by the steam flow. Similarly positive results had been reported by Masuda et al. [247], who proposed a new system with the catalysts $FeO(OH)$ (hydrated ferric oxide) and Ni supported on a rare-earth metal exchanged-Y zeolite (Ni-REY), consisting of three reactors connected in series. Among eight plastics (polyurethane, two polyolefins, two polyamides and three polyesters), it was found that only in the case of the polyesters (PC, PET and PBT) did the introduction of steam affect the reaction mechanism; besides this, the amount of carbonaceous residue was negligible in this system.

Another interesting approach was developed by Kaminsky et al. [245], who had to deal with the real problem of applying pyrolysis to poly (methyl methacrylate) (PMMA) carrying a big load of alumina trihydrate (ATH), a filler used at concentrations of up to 67% as a fire retardant. The aim was to study the influence of this additive on the pyrolysis of PMMA. In fluidized-bed reactors operating at 450 °C, the decomposition of PMMA attained a monomer yield of up to 97.2 wt%. In the presence of ATH at 450° C, filled PMMA yielded only 58% MMA monomer and not only did the ATH present no catalytic effect but also water formed by its dehydration was found to lower the monomer yield.

Chapter 6

ENERGY RECYCLING

Energy recycling differs from the recovery of fuels by thermolysis in the fact that residues are incinerated at temperatures above 1000°C, so that their energy content is used directly for heating or to generate steam and electric power. The calorific value varies with the composition of the residue. Typical values for waste materials are shown in table 9. It can be seen that the amount of energy recovered from plastics is, in general, among the highest, the polyolefins attaining the level of fuel oil [248-251].

**Table 9. Heating values of some combustible
materials and residues in MSW [248]**

Materials, Residues	Heating values (kcal/kg)	Materials, Residues	Heating values (kcal/kg)
Polyethylene	18,720	Newspapers	6,970
Polypropylene	18,434	Pine wood*	9,100*
Polystyrene	16,082	Vegetable oils	14,770
Tire chips*	13,000-14,000*	Bituminous coal*	11,000-14,000*
Polyvinyl chloride	7,516	Fuel oil*	18,000-19000*
Polyurethane foam	11,362	Food waste	4,200
Nylon	10,138	Textiles	5,900
Phenol formaldehyde	13,197	Corrugated boxes	6,153

* Data from [249], in BTU/lb.

During the 1990s, there was a lot of controversy about energy recycling, generated chiefly by an argument between the industry and environmentalists, concerning the emission of gases, release of dioxins and disposal of solid waste.

While this argument was going on, at the end of the eighties and start of the nineties, several countries of the EU, North America and Japan sent most of their waste to "incineration facilities", not all of which had installed energy-recovery equipment. In fact, in some facilities the refuse was simply burnt, to reduce its volume or to destroy pathogenic material, without any care being taken to avoid toxic and polluting emissions [248, 250].

In that period, if combustors both with and without energy recycling are considered, more than 40% of all MSW was incinerated in France (42%), Sweden (47%), Denmark (48%), Belgium (54%) and Switzerland (59%), and as much as 75% in Luxemburg. On the other hand, in the USA there was a marked fall in incinerated MSW, from 31% in 1960 to little more than 7% in 1988. However, according to projected trends published by Alexander [250], the number of 'waste-to-energy' plants in the USA would rise from 128, operating in 1990, to 188 in the late nineties. With this program, 37% of all residues would be incinerated, saving the equivalent of 57 million barrels of gasoline per annum.

Already in the early nineties, almost every Japanese community operated a waste-to-energy facility, totaling 70% of garbage incinerated, even though their recycling rate was more than double that of the USA. On the other hand, in Canada in 1987, where domestic waste per capita was the highest in the world (1.7kg/day), a large part of this was buried and only a tiny amount burnt in the mere twelve incinerators possessed by the country. Maraghi argued in favor of using energy from waste in Canada, on the basis of his review of the subject [248].

Even in countries not described as developed, specific regional problems may motivate big investments in building combustion facilities. A typical example was mentioned by Abu-Hijleh and co-workers [251], who concluded that it would be economically feasible to build a waste incineration plant in Jordan at the end of the nineties. They used sensitivity analysis to investigate the profitability of a material recycling and electricity generation plant under different financial scenarios. The need for such a waste disposal alternative at that time was urgent, because of a substantial increase in household and commercial waste generation in Jordan, resulting from a rapid growth in the Gross National Product (GNP) of 6% simultaneously to a high population growth rate (3.6%) and increasing standard of living.

Another typical example of specific local factors is provided by Sweden, a fully-developed country whose towns are having problems complying with the new waste management laws. According to Holmgren and Henning [252], these stipulated that, from 2005, combustible and organic residues were to be banned from landfills, and a new tax was imposed on landfill disposal of waste. This situation provoked plans for new combustion plants which, if carried out, would almost double the waste incineration capacity of these towns from 2.4 to 4.7 million tons a year, between 1998 and 2006.

From the environmental viewpoint, the greenhouses gases, dioxins and other organic compounds, as well as heavy metals, generated by these incinerators are always seen as arguments against energy recycling. From the technical and economic point of view, the main obstacles are the need to separate food and organic garbage from non-combustible components, so as to obtain the highest possible energy harvest, and the massive investments demanded for new installations. On the other hand, the very important points in favor of this method of recovering waste are the drastic reduction in the volume of buried refuse and the savings in non-renewable resources that would otherwise be burned to generate heat and power [248-253].

According to certain authors and the power industry, the problem of pollution by emissions and effluent from incinerator plants was virtually solved a long time ago. Dangerous particulates such as fly ash and low-volatility metal vapors are retained in bag-filters and electrostatic precipitators. Acid gases like HCl and SO_2 are neutralized with lime or caustic soda. On the question of the impact on global warming, the controlled incineration of a ton of residues in combustors liberates 6 to 10-fold less total greenhouse gas than does the same quantity decomposing in a landfill. In addition, the oxides of nitrogen (NO_x) produced by combustion of residues are up to 250 times more efficient per molecule than CO_2 in trapping heat, and they can be reduced in combustors to N_xO, by adjusting the amount of air in the combustion mixture or spraying with ammonia solution [248, 250, 253].

Regarding dioxins, without entering the unresolved argument about whether or not a real toxic risk to human health has been demonstrated in atmospheric concentrations of the order of parts per trillion or quadrillion, modern incinerators are designed to operate at a temperature between 925 and 1,315°C, to eliminate all dioxin emissions. This is done because the dioxins present in the residues are destroyed above 925°C, while those produced by the combustion do not form below 1,315°C [250].

Another controversy surrounds the heavy metals which are actually present in the ashes. The ashes are generated directly as the solid residue of combustion (bottom ash), with a quite limited potential for pollution, and also as particles removed from exhaust gases in the flue (fly ash) [248, 250, 253], which are heavily loaded with these metals. Heavy metals can be leached from these ashes, when they are inadequately buried in landfills, eventually contaminating the ground water.

One of the cited authors [250] comments that the most important heavy metals, lead and cadmium, are leached from ash in laboratory tests in strongly acid solutions, whereas in practical industrial analyses, the ash residues are solely buried in the mildly alkaline conditions produced by the ash itself, resulting in a leachate that remains free of heavy metals even after 20 years of monitoring. Moreover, given that most of the lead and cadmium in MSW comes from discarded batteries, the mere separation of these items from the residues before incineration would substantially reduce the heavy-metal content at source and thus in the ashes.

Nonetheless, Huang *et al.,* [253] point out that, in order to comply with current local regulations such as those in, say, Taiwan, fly ash destined for sanitary landfills must first undergo intermediate treatments such as solidification, stabilization or others, specified in the criteria of the Taiwan Environmental Protection Administration (TEPA). Unfortunately, the explosive rise in the volume of refuse generated on that island-country, coupled with the lack of space, has led to the predominance of incineration of MSW over other disposal methods. Already in 2003, 70% by weight of all MSW in Taiwan was incinerated in 19 facilities, although the total capacity of the incinerators does not represent the whole potential capacity for energy recovery. Around 2007, all towns will be served by combustion plants, worsening the problem of what to do with the ash produced, which amounts to more than 18% of the incinerated MSW and already in 2003 totaled 1.05 million tons.

Recently, there has been a focus on comparative evaluation of energy recovery and material recovery processes. Patel and Xanthos [4] have presented a summary of results obtained in earlier work, in which they used a method based on the principles of life-cycle analysis (LCA) to estimate the energy consumed in the various possible pathways followed by plastic residues. The values derived from the analysis of energy flow were converted to figures on a scale of merit, on which waste-to-energy incineration scores 100. For each kind of waste disposal, material and energy recovery were ranked in terms of energy-saving efficiency, values higher than 100 representing processes less efficient than waste-to-energy combustion of plastics. The authors concluded that landfill, where obviously all the energy content of the waste is lost, and thermolysis to fuels are less efficient than energy recovery, while, in the following order, thermolysis to monomers < mechanical recycling < reuse of articles are more efficient.

Holmgren, Henning and co-workers have published several articles on this topic. In some of the most recent, they described important results obtained with the aid of the energy system optimization model MODEST (Model for Optimization of Dynamic Energy System with Time-dependent components and boundary conditions). Apart from being used to review the various studies made of the Linköping utility [254], MODEST was used for a comparative assessment of material and energy recovery from MSW in the energy utilities of two Swedish towns: Linköping, which at the end of 2003 already used several energy-sources, the basic heat supply coming from the incineration plant, and Skövde, which used a heat plant fired by wood chip and was planning to invest in the construction of a waste-incineration plant [252].

Holmgren and Henning concluded that investing in plants using waste as a fuel for electricity and heat production would be profitable for the two towns. However, non-combustible residues, such as metals and glass, give no heat contribution when incinerated; material recycling of metals saves a lot of energy, whereas glass material recycling saves less. Moreover, considering the energy efficiency calculation, waste-to-energy is preferable for cardboard and biodegradable waste, whereas material recovery should be applied to paper and hard plastics, even if a district heating system is able to use the energy generated by a waste of such high combustion energy [252, 254]. This conclusion for plastics is in agreement with the data presented by Patel and Xanthos that ranked energy savings of the different forms of plastics waste recovery [4].

In Brazil, not a single unit exists that produces energy from MSW. There was practically no discussion over alternative sources of power and heating, until this question began to acquire importance in 2001, with a power crisis provoked by a shortage in the supply of electricity. The government had to subject industry, commerce and the general public to a drastic rationing of power consumption. In November 2005, a technical cooperation agreement was signed between Federal Government companies (Eletrobrás and the Electric Power Thermal Generation Company – CGTEE) and the cleansing department of the municipal government of Porto Alegre in Rio Grande do Sul (DMLU), to promote the production of gas from the biodigestion of MSW. The biogas produced will be used to generate electric power at the Nutepa power Station of Porto Alegre [255]. We have no knowledge of any other plans for the construction of this type of plant or of a waste-to-energy incineration facility in the country.

Chapter 7

REMAINING CHALLENGES

While working on this chapter, the authors came to realize how much effort has been and is being invested by researchers and government bodies, enterprises and, not least, popular initiatives around the world, in attempts to make all stages of plastic waste recycling viable from the technical, economic and social points of view. It was also clear, though, that there is still a lot more to be done if the recovery of plastic residues is to make an effective contribution to reaching the goal of sustainable development in the strict sense. Thus, to end this chapter, the authors thought it worthy to mention some of the challenges in plastic recycling that still have to be overcome.

The first is to increase the rate of material recovery, principally by mechanical recycling which, in terms of energy savings, is the most suitable form of plastics recycling in most cases [4, 256]. To achieve this there will need to be more national laws and international agreements/directives establishing targets and allocating responsibility for the final destination of residues mainly to the manufacturers, while not exempting the public at large from playing its important part. The compulsory taxation on non-compliance with these regulations may seem a rather imposed incentive, but it would be very effective, even though in high population countries of continental proportions, like the USA, Russia, China, Brazil and India, the difficulty of regular inspection would have to be considered.

Nowadays, besides the short-life packaging, there has been an alarming rise in the generated volume of plastic residues from engineering, from scrapped vehicles and electro-electronic products such as computers, mobile phones and photocopiers, which accumulates year by year in a growing mountain of plastic. Often, these components are made out of polymer blends or multi-material systems that are hard to separate into non-contaminated homogeneous materials, or out of laminates produced by *in-situ* formation of the polymer matrix, as is the case in polyester or polyamide mats with a polyolefin-blend backing layer, which are not economically viable to be recycled separately using the current technologies [42, 122, 257-259].

On the other hand, Massura and co-workers [260] recently tested in a pilot plant a technique to separate the layers of multilayer flexible packages, typically constituted by the following components: PET/inks/adhesive/aluminum/adhesive/PE. They achieved an efficient separation by using specific solvents to dissolve the adhesives and inks, frequently based on polyurethanes, and even put the separate components on the market. However, the enterprise could not advance, due to lack of interest on the part of national investors from this market sector in Brazil.

Within the stream of plastic waste there are some residues considered harmful, such as containers used for toxic agrochemicals and pathogenic plastic residues from hospitals [261], and some that are not classed as dangerous, but are included in producer-responsibility recycling systems [262] or are banned from landfills by government regulations [256], such as combustible organic residues. There is an urgent need, therefore, to find solutions to this problem, creating suitable treatment facilities or destinations for these materials that cannot be buried.

In the case of the agrochemical containers, apart from being left on the land without due care, they are commonly reused for the transport and storage of water and food by the landowners and rural workers, who frequently are ignorant of their polluting and toxic potential. Generally, after the toxic contents have been fully used, about 0.3% of the initial volume remains in the container and, following triple washing in clean water with a pressurized jet, 99.9% of that volume is removed, reducing considerably the problem of finding a suitable destination [263]. These residues that have already been the target of Brazil's government, nowadays have their own market and recycling facilities established, due to their available collection system achieved through the enforced government regulations. From a total of approximately 14,000 tons of plastic agrochemical containers collected in these facilities, more than 50% were successfully recovered in 2004, mainly through mechanical recycling to manufacture products for the construction sector.

One great challenge is the recycling of commodity plastics, including the various types of PET bottles, into high added-value products. These residues, in contrast to the engineering plastics, are of low value and generated in enormous quantities, and thus demand recovery facilities on a massive scale. We take the view that material recovery should be adopted in this case, employing both chemical processes for the production of monomers and mechanical recycling to produce valuable articles, such as the previously mentioned closed-loop bottle-to-bottle product and ecological synthetic paper.

As far as this synthetic paper is concerned, the forthcoming pilot-plant tests with recycled plastic residues should tell whether or not the present technique is viable for commercial application. On the commercial side, the Brazilian suppliers of synthetic paper are unanimous in the view that the substitution of recycled plastic for virgin resin in the same process to make multilayer films would afford substantial financial and environmental benefits.

In the case of PET, PMMA, PS and PTFE, the industrial exploitation of depolymerization is well-established [264]. By contrast, selective depolymerization of polyolefins to generate high yields of monomers remains technically an unsolved challenge. High-yield recycling processes and mass production are needed for polyolefins, as they are the class of plastics found in greatest amounts in waste. In addition, although the method of pyrolysis of these residues is well-established, the various products derived from the reaction are determined mainly by secondary reactions, causing difficulties of control in production. Kaminsky and Hartman [264] have reported a process to degrade PE which they describe as particularly spectacular. According to these authors, the method was invented by Dufaud and Basset, who utilized a strongly electrophilic Ziegler-Natta catalyst, zirconium monohydride, at 150°C, resulting in the complete decomposition of the polymer into ethane and methane after 15h of reaction.

Kaminsky and Hartman [264] also comment that this process is a random chain-cleavage reaction, which should not be considered a true depolymerization or retropolymerization. According to them, given that this is the first case of a catalyst actively involved in the primary reaction, Dufaud and Basset have made a remarkable discovery. However, it was still not possible to obtain the ethylene monomer and, while they consider that "catalytic monomer recovery is a large step closer", this remains a challenge for the chemical recycling of polyolefins.

The authors realize that sustainable development in plastics recycling will not be fully attained if only the ongoing scenario is to be continued, but also are optimistic in contributing to the needed development of economically viable and clean recycling technologies or even to the improvement of currently available recycling technologies. Furthermore, the authors realize that financial support from the private sector and the government enforcement regulations, together with public participation, are crucial to promote the recovery of the highest added value of recycled plastics before plastics become scarce and much of the valuable energy of their residues has been exhausted.

ACKNOWLEDGMENTS

The authors thank to São Paulo State Foundation for Research Support (FAPESP), PADCTIII/CNPq, CAPES, Polibrasil Resinas SA for the financial support and students scholarships. The authors are also grateful to Eliton Souto de Medeiros and Silvio Manrich for their valuable comments and special aid.

REFERENCES

[1] Ross, S.; Evans, D. *J. Cleaner Production.* 2003, 11, 561-571.

[2] Maeda, M.; Ikeda, T. *Polym. Adv. Technol.* 2000, 11, 388-391.

[3] (2005). Canada faces climate crisis. *Carbon Brazil,* http://www. carbonobrasil.com (in Portuguese).

[4] Patel, S.H.; Xanthos, M. *Adv. Polym. Technol.* 2001, 20(1), 22-41.

[5] Pagliarussi, M. S. M.S. thesis, Federal University of São Carlos, São Carlos, SP, 1995 (in Portuguese).

[6] Risso, W.M.; Wiebeck, H. *Plastics recycling and their industrial applications,* University of São Paulo, São Paulo, SP, 1994 (in Portuguese).

[7] Amaral, L.H. *Folha de São Paulo.* September 16, 1994 (in Portuguese).

[8] Fridman, I.J. *Rev Limp Públ.* 1997, 44, 11-12 (in Portuguese).

[9] Duchin, F.; Lange, G. M. *Struct Change Econ Dynam.* 1998, 9, 307-331.

[10] Agnelli, J. A. M., *et al. Polím: Ciência e Tecnol.* October/December, 1996, 9-18 (in Portuguese).

[11] Landsberger, S.; Chichestes, D. L. *J. Radioanal Nucl. Chemistry.* 1995, 192, 289-297.

[12] Mancini, S.D., Zanin, M. In: *Congress of Sanitary Engineering,* Peru, 1998, 27th.

[13] Dent, I. *Plast Packaging Recy.* 1999.

[14] *Warmer Bull.* January 1997, 52.

[15] Broek, W. H. A. M., *et al. Chemometr. Intell. Lab. Systems.* 1996, 35, 187-197.

[16] Tombs, G. *Plast. Packaging Recy.,* 1999.

[17] Toloken, S. *Plást. Rev.* November 2002, 38 (in Portuguese).

[18] Borges, A. *Gazeta Mercantil.* March 24, 1999, C-6 (in Portuguese).

[19] *Plást. Rev.* June 2001, 28 (in Portuguese).

[20] Gimenez, K. *Plást. Rev.* May 2002, 22 (in Portuguese).
[21] Davis, G.; Phillips, P. S.; Read, A. D.; Lida, Y. *Resour. Conserv. Recy.*, 2006, 46, 115-127.
[22] Daskalopoulos, E.; Badr, O.; Probert, S. D. *Appl Energ.* 1997, 58, 209-255.
[23] Onusseit, H. In: *AFERA Technical Seminar*, Düsseldorf, 2002, 1st.
[24] Mancini. S.; Santos, A. S. F.; Manrich, S.; Zanin, M. *Pack.* 2001, 46, 36-37 (in Portuguese).
[25] *An Analysis of Plastics Consumption and Recovery in Europe.* Plastics Europe Info Point: Bruxels, 2004.
[26] Dennison, M. T.; Mennicken, T. In: *PACIA CONVENTION*, Brisbane, 1996.
[27] Tawfik, M. S.; Devlieghere, F.; Steurbaut, W.; Huyghebaert, A. *Acta Alimentaria.* 1997, 26, 219-233.
[28] Tilman, C.; Sandhu, R. *Resour Conserv. Recy.* 1998, 24, 183-190.
[29] Esposito, F. *Plást. Rev*, May 2001, 28 (in Portuguese).
[30] Toloken, S. *Plást. Rev*, September 2002, 33 (in Portuguese).
[31] *Napcor Reports Increase in PET Recycling Rate and Record Volume of Recycled PET Containers.* NAPCOR: Sonoma, CA, 2005.
[32] *Performance and perspectives of plastic's recycling in Brazil.* PLASTIVIDA: São Paulo, SP, 2004 (in Portuguese).
[33] Mancini, S. D.; Zanin, M. *Plást. Ind.* September 2000, 118.
[34] *Plastic's recycling in Brazil.* CEMPRE: São Paulo, SP, 2004 (in Portuguese).
[35] *Embanews.* January 2002, 46.
[36] Lagarinhos, C. A F. M.S. thesis. Institute of Technological Research, São Paulo, SP, 2004 (in Portuguese).
[37] Camponero, J. PhD. thesis. University of São Paulo, São Paulo, SP, 2002 (in Portuguese).
[38] Resolution n. 258. Ministry of Environment, National Council of Environment, Brazil, August 26, 1999 (in Portuguese).
[39] Manrich, S.; Frattini, G.; Rosalini, A. C. *Plastics identification: a recycling tool.* Editora da UFSCar (EdUFSCar): São Carlos, SP, Brazil, 1997; 23-42 (in Portuguese).
[40] Kukaleva, N.; Simon, G.P.; Kosior, E. *Polym. Eng. Sci.* 2003, 43, 431-443.
[41] Massura, A. C. M.S. thesis. University of São Carlos, São Carlos, SP, Brazil, 2005 (in Portuguese).
[42] Tsunekawa, M.; Naoi, B; Ogawa, S.; Hori, K.; Hiroyoshi, N.; Ito, M.; Hirajima, T. *Int. J. Miner Process.* 2005, 76, 67-74.
[43] Pascoe, R. D. *Waste Manage.* 2006 (in press).

[44] SRH Simco/Ramic catalog. Simco/Ramic Corporation, Inc. 1992.
[45] Collier, M. C.; Baird, D. G. *Polym. Comp.* 1999, 20, 423-435.
[46] Van Ness, K. E.; Nosker, T. J. In: *Plastics recycling: products and processes*; Ehrig, R. J.; Ed.; Hanser Publishers: Munich, 1992, 187-229.
[47] Nauf, U.; Mäurer, A.; Holley, W.; Wiese, M.; Utschick, H. *Kunststoffe.* 2000, 90.
[48] Drelich, J.; Payne, T; Kim, J. H.; Miller, J. D.; Kobler, R.; Christiansen, S. *Polym. Eng. Sci.* 1998, 38, 1378-1386.
[49] Marques, G. A.; Tenório, J. A. S. *Waste Manage.* 1999, 20, 265-269.
[50] Hearn, G. L.; Ballard, J. R. *Resour. Conserv. Recy.* 2005, 44, 91-98.
[51] Green, J. L.; Petty, C. A.; Gillis, P. P.; Grulke, E. A. *Polym. Eng. Sci.* 1998, 38, 194-203.
[52] Saito, T.; Satoh, I. *Polym. Eng. Sci.* 2005, 1419-1425.
[53] Huber, M., Franz, R. *J. High Resol. Chromatogr.* 1997, 20, 427-430.
[54] Syrinek, A. R.; Tex, R. US Patent 5,330,581, 1994.
[55] Sanko, G. Recycling product news, july/august 1999, 7.
[56] Kao, C., Cheng, W., Wan, B. *J. Appl. Polym. Sci.* 1998, 70, 1939-1945.
[57] Santos, A. S. F.; Teixeira, B. A. N.; Agnelli, J. A. M.; Manrich, S. *Resour. Conserv. Recy.* 2005, 45, 159-171.
[58] Heyde, M.; Kremer, M. *Recycling and recovery of plastics from packaging in domestic waste.* Ecomed Publishers: Freising, 1999, 5.
[59] Santos, A. S. F.; Araújo, E. S.; Manrich, S. In: *Brazilian Polymer Congress*, Associação Brasileira de Polímeros: São Carlos, SP, 1999, 5th (in Portuguese).
[60] Remédio, M. V. P. M.S. thesis, Federal University of São Carlos, São Carlos, SP, 1999 (in Portuguese).
[61] Frattini, G. M.S. thesis, Federal University of São Carlos, São Carlos, SP, 1999 (in Portuguese).
[62] Santos, A. S. F. M. S. thesis, Federal University of São Carlos, São Carlos, SP, 2000 (in Portuguese).
[63] Oldshue, J. Y. *Chem. Eng.* June 1983, 90, 82-108.
[64] Rauwendaal, C. *Polymer Extrusion.* Hanser Publishers: Munich, 2001.
[65] Strumillo, C.; Kudra, T. *Drying: principles, applications and design.* Gordon and Breach Science Publishers: Montreux, 1986.
[66] Porter, H. F. *et al.* In: *Perry's Chemical Engineers Handbook.* Perry, R. H. (ed.). 6. ed. Mc Graw-Hill: New York, 1984, 20-21.
[67] Jomaa, W.; Baixeras, O. *Drying Technol.* 1997, 15, 2129-2144.
[68] Piovan S.p.A. *Piovan PET systems.* Venice, 1997.

[69] Laoutid, F.; Ferry, L; Lopez-Cuesta, J. M.; Crespy, A. *Fire Mater.* 2006 *(in press).*

[70] Greco, A.; Maffezzoli, A.; Manni, O. *Polym. Degrad. Stab.* 2005, 90, 256-263.

[71] Japon, S.; Leterrier, Y.; Manson, J-A. E. *Polym. Eng. Sci.* 2000, 40, 1942-1952.

[72] Poulakis, J. G.; Papaspyrides, C. D. *Advan. Polym. Technol.* 2000, 19, 203-309.

[73] Papaspyrides, C. D; Kartalis C. N. *Polym. Eng. Sci.* 2000, 40, 979-984.

[74] Shin, C.; Chase, G. G.; Reneker, D. H. *Colloid Surface.* 2005, 262, 211-215.

[75] Xanthos, M.; Dey, S. K.; Mitra, S.; Yilmazer, U. Feng, C. *Polym. Comp.* 2002, 23, 153-163.

[76] González, O.; Muñoz, M. E.; Santamaría, A.; García-Morales, M.; Navarro, F. J.; Partal, P. *Eur. Polym. J.* 2004, 40, 2365-2372.

[77] Henninger, F.; Drake, W. O.; Sitek, F. *The role of processing stabilizers in recycling of polyolefins.* Basel: Ciba-Geigy.

[78] Agnelli, J. A. M. *Degradation, stabilization and aging of polymers.* Associação Brasileira de Polímeros (ABPol): São Carlos, SP, 1997, 196-204 (in Portuguese).

[79] Popisil, J. *Int. Polym. Sci. Technol.* 1994, 21, T/54-T/58.

[80] Paci, M.; La Mantia, F. P. *Polym. Degrad. Stab.* 1998, 61, 417-420.

[81] Villain, F., Coudane, J., Vert, M. *Polym. Degrad. Stab.* 1995, 49, 393-397.

[82] Allen, N. S; Edge, M; Daniels, J; Royall, D. *Polym. Degrad. Stab.* 1998, 62, 373-383.

[83] Ravindranath, K; Mashelkar, R.A. *Chem. Eng. Sci.* 1986, 41, 2197-2214.

[84] Edge, M; Allen, N.S; Wiles, R; McDonald, W; Mortlock S.V. *Polym.* 1995, 36, 227-234.

[85] Edge, M; Allen, N.S; Wiles, R; McDonald, W; Mortlock S.V. *Polym. Degrad. Stab.* 1996, 53, 141-151.

[86] Santos, A. S. F.; Medeiros, E. S.; Agnelli, J. A. M.; Manrich, S. In: *World Polymer Congress*, IUPAC: Paris, 2004.

[87] Paul, D. R., Keskkula, H. *Encyclopedia of the polymer science and engineering.* John Wiley: New York, 1986, 4, 630-691.

[88] Foster G. N.; Wasserman S. H.; Yacka D. J. *Die Angew Makromol. Chem.* 1997, 252, 11-32.

[89] Launay, A.; Thominette, F.; Verdu, J. *Polym. Degrad Stabil.* 1999, 63, 385-389.

[90] Agnelli, J.A.M., Chinelatto, E. M. A. *Polím: Ciência Tecnol* July-/September 1992, 127-131.

[91] Allen, N.S., *et al. Polym. Degrad. Stab.* 1994, 43, 229-237.
[92] Zaikov, G. E. *Int. Polym. Sci. Technol.* 1991, 18, T/49-T/59.
[93] Cardi, N., *et al. J. Appl. Polym. Sci.* 1993, 50, 1501-1509.
[94] Ballara, A.; Verdu, *J. Polym. Degrad. Stab.* 1989, 26, 361-374.
[95] Hendrickson, L.; Connole, K. B. *Polym. Eng Sci.* 1995, 35, 211-217.
[96] Zimmerman, H.; Kim, N. T. *Polym. Eng. Sci.* 1980, 20, 680-683.
[97] Craig, I. H.; White, J. R.; Kin, P. C. *Polym.* 2005, 46, 505-512.
[98] Agnelli, J. A. M., Sousa, J. A., Canevarolo Jr., S. V. *Introduction to degradation and stabilization of polymers.* Associação Brasileira de Polímeros (ABPol): São Carlos, SP, 1991, 24-30 (in Portuguese).
[99] Pospísil, J. *Polym. Degrad. Stab.* 1993, 40, 217-232.
[100] Neri, C. *et al. Polym. Degrad. Stab.* 1995, 49, 65-69.
[101] Henninger, F.; Gugumus, F.; Pedrazzetti, E. *Processing, heat and light stabilization of polyolefins.* Buenos Aires: Ciba-Geigy, 1986.
[102] Gugumus, F. *Polym. Degrad. Stab.* 1994, 44, 273-297.
[103] Monteiro, M.; Nerín, C.; Reyes, G. R. *Food Addit. Contam.* 1996, 13, 575-586.
[104] Hesbst, H. *et al. Restabilization produces high quality recycled polyolefins.* Munique: Carl Hanser Verlag, 1992.
[105] Santos, A. S. F.; Agnelli, J. A. M.; Trevisan, D. W.; Manrich, S. *Polym. Degrad. Stab.* 2002, 77, 441-447.
[106] Herbst, H. et al. In: International Recycling Congress at the Re'93 Trade Fair. Palexpo: Genebra, 1993, 3, 332-34.
[107] Angrawal, A. K.; Singh, S. K.; Utreja, A. *J. Appl. Polym. Sci.* 2004, 92, 3247-3251.
[108] Malík, J. *et al. Polym. Degrad. Stab.* 1995, 50, 329-336.
[109] Awaja, F.; Pavel, D. *Eur. Polym. J.* 2005, 41, 1453-1477.
[110] Pacheco, E. V; Dias, M. L. *J. Polym. Eng.* 2003, 23, 23-42.
[111] Li, J.; Favis, B. D. *Polym.* 2001, 42, 5047-5053.
[112] Zumbrunnen, D.A.; Chhibber, C. *Polym.* 2002, 43, 3267-3277.
[113] Hage Jr., E.; Ferreira, L.A.S.; Manrich, S.; Pessan, L.A. *J. Appl. Polym. Sci.* 1999, 71, 423-430.
[114] Lyngaae-Jorgensen, J.; Utracki, L.A. *Polym.* 2003, 44, 1661-1669.
[115] Willemse, R.C.; Speijer, A.; Langeraar, A.E.; Boer, P. *Polym.* 1999, 40, 6645-6650.
[116] Halimatudahliana, H.I.; Nasir, M. *Polym. Test.* 2002, 21, 163-170.
[117] Wu, J.; Guo, B.; Chan, C-M.; Li, J.; Tang, H-S. *Polym.* 2001, 42, 8857-8865.
[118] Danella Jr, O.J.; Manrich, S. *Polym. Sci.* 2003, 45, 1086-1092.

[119] Fortelny, I.; Michálková, D.; Mikesová, J. *J. Appl. Polym. Sci.* 1996, 59, 155-164.

[120] Cherian, Z.; Lehman, R. *Int J Adhesion Adhes.* 2005, 25, 502-506.

[121] Bonelli, C.M.C.; Martins, A.F.; Mano, E.B.; Beatty, C.L. *J. Appl. Polym. Sci.* 2001, 80, 1305-1311.

[122] Balart, R.; López, J.; García, D.; Salvador, M.D. *Eur. Polym. J.* 2005, 41, 2150-2160.

[123] Liu, X.; Boldizar, A.; Rigdhal, M.; Bertilsson, H. *J. Appl. Polym. Sci.* 2002, 86, 2535-2543.

[124] Elmaghor, F.; Zhang, L.; Li, H. *J. Appl. Polym. Sci.* 2003, 88, 2756-2762.

[125] Fraïsse, F.; Verner, V.; Commereuc, S.; Obadal, M. *Polym. Degrad. Stab.* 2005, 90, 250-255.

[126] Pawlak, A.; Morawiek, J.; Prazzagli, F.; Pracella, M.; Galeski, A. *J. Appl. Polym. Sci.* 2002, 86, 1473-1485.

[127] Yu, Z-Z.; Yang, M-S.; Dai, S-C., Mai Y-W. *J. Appl. Polym. Sci.* 2004, 93, 1462-1472.

[128] Santana, R. M. C.; Manrich, S. *J. Appl. Polym. Sci.* 2003, 87, 747-751.

[129] Manrich, S; Sanches, L. F.; Perez, I. S. B. Federal University of São Carlos, São Carlos, SP (unpublished).

[130] Corrêa, A. C.; Manrich, S. In: *International Macromolecular Colloquium*, Gramado, RS, 2005, 10th.

[131] Ravazi, R. F. M.S. thesis, University of São Carlos, São Carlos, SP, Brazil, 2002 (in Portuguese).

[132] Lima, C. R.; Hage, E.; Manrich, S. In: *Conference of the Brazilian Society of Metals and Materials*, Águas de Lindóia, SP, 2001 (in Portuguese).

[133] Danella Jr, O. J.; Manrich, S. Federal University of São Carlos, São Carlos, SP (unpublished).

[134] Manrich, S.; Radanitsch, J.; Manrich, S.; Almeida, M.C. In: *Brazilian Polymer Congress*, Associação Brasileira de Polímeros: São Carlos, SP, 2003, 7th (in Portuguese).

[135] Navrátilová, E.; Fortelný, I. *Polym. Network. Blend.* 1996, 6, 127-133.

[136] Scaffaro, R.; La Mantia, F. P. *Polym. Eng. Sci.* 2002, 42, 2412-2417.

[137] Scaffaro, R.; Dintcheva, N.T.; Nocilla, N.A.; La Mantia, F.P. *Polym. Degrad. Stab.* 2005, 90, 281-287.

[138] Mallick, P. K.; Newman, S. *Composites materials technology: processes and properties.* Hanser Publishers: Munich, 1990.

[139] Chawla, K. K. *Composite materials: science and engineering.* 2nd Ed. Springer-Verlag: New York, 1998.

[140] White, R. J.; De, S. K. Short-fiber polymer composites. Woodhead Publishing:Cambridge, 1996.

[141] Manson, J.A.; Sperling, L.H. *Polymer Blends and Composites*. Plenum Press: New York, 1976.

[142] Liang J. Z., Tang C. Y., Li, R. K. R., Wong T. T. *Metal. Mater.* 1998, 4, 616-619.

[143] Thio, Y.S.; Argon, A.S.; Cohen, R.E.; Weinberg, M. *Polym.* 2002, 43, 3661-3674.

[144] Silva, A. L. N.; Rocha, M. C. G.; Moraes, M. A. R.; Valente, C. A. R.; Coutinho, F. M. B. *Polym. Test.* 2002, 21, 57-60.

[145] Nielsen, L. E. *Mechanical Properties of Polymers and Composites*, Marcel Dekker: New York, 2, 1974.

[146] Medeiros, E. S.; Agnelli, J. A. M.; Joseph, K.; Carvalho, L. H.; Mattoso, L. H. C. *Polym. Comp.* 2005, 26, 01-11.

[147] Jiang, H.; Kamdem, D. P. *J. Vinyl. Addit. Technol.* 2004, 10, 59-69.

[148] Arbelaiz, A. *et al. Comp Sci Technol.* 2005, 65, 1582-1592.

[149] Wu, Y.; Isarov, A. V.; O'Connell, C. *Thermochim. Acta.* 1999, 340-341, 205-220.

[150] Guffey, V. O.; Sabbagh, A B. *J. Vinyl. Addit. Technol.* 2002, 8, 259-263.

[151] Ha, C. S.; Park, H. D.; Cho, W. J. *J Appl. Polym. Sci.* 1999, 74, 1531-1538.

[152] Dintcheva, N. T.; La Mantia, F. P. *Polym. Advan. Technol.* 1999, 10, 607-614.

[153] Santos, P.; Pezzin. S. H. *J. Mater Process Technol.* 2003, 143-144, 517-520.

[154] Kouparitsas, C. E.; Kartalis, C. N.; Varelidis, P. C.; Tsenoglou, C. J.; Papaspyrides, C. D. *Polym. Comp.* 2002, 23, 682-689.

[155] Liu, Y. ; Meng, L ; Huang, Y. ; Du, J. *J. Appl. Polym. Sci.* 2004, 95, 1912-1916.

[156] Licea-Claveríe, A.; Carrillo, F. J. U.; Alvarez-Castillo, A.; Castaño, V. M. *Polym. Comp.* 1998, 20, 314-320.

[157] Pannkoke, K.; Oethe, M.; Busse, J. *Cryogen.* 1998, 38, 155-159.

[158] Rothon, R.N.; Hancock, M. *Particulate-Filled Polymer Composites*. Longman Scientific & Technical; England, 1995.

[159] Tadmor, Z.; Gogos, C. G. *Principles of Polymer Processing*. John Willey & Sons, Inc.: New York, 1979.

[160] Pierce, D. E., Pfeffer, R. L., Sadler, G. D. *Nucl. Instrum. Meth. Phys. Res. B.* 1997, 124, 575-578.

[161] Bayer, F. L. *Food Addit. Contam.* 1997, 14, 661-670.

[162] Food and Drug Administration. (2003). *Points to consider for the use of recycled Plastics in Food Packaging: Chemistry Considerations.* http://www.fda.gov

[163] Devlieghere, F.; Huyghebaert, A. *Lebensm-Wiss Technol.* 1997, 30, 62-69.

[164] Kuznesof, P. M.; Vanderveer, M. C. In: *Plastics, Rubber, and Paper Recycling.* American Chemical Society: New York, 1995, 389-403.

[165] Begley, T. H. *Food Addit. Contam.* 1997, 14, 545-553.

[166] Code of Federal Regulations 58, Food and Drug Administration, United States of America, 1993, 52719.

[167] Code of Federal Regulations 60, Food and Drug Administration, United States of America, 1995, 36582.

[168] International Life Sciences Institute. *Guidelines for recycling of plastics for food contact use.* Brussels, 1997.

[169] Bündesinstitut für Gesundheitlichen Verbraucherschutz und Veterinärmdizin. *Use of mechanically recycled plastic made from polyethylene terephthalate (PET) for the manufacture of articles coming into contact with food.* 2001.

[170] Plastics Recycling Task Force; National Food Processors Association; The Society Of The Plastics Industry, Inc. *Guidelines for the safe use of recycled plastics for food packaging applications.* 1995.

[171] Franz, R., Huber, M., Welle, F. *Deut Lebensm-Rundsch.* 1998, 94, 303-308.

[172] Nielsen, T.; Damant, A. P.; Castle, L. *Food Addit. Contam.* 1997, 14, 685-693.

[173] Welle, F; Franz, R. Maack Business Serv. 1999, Session XII/3, 1-8.

[174] Komolprasert, V.; Lawson, A. In: *Annual Technical Conference.* Society of Plastics Engineers, 1994. 2906-2908, 52nd.

[175] Demertzis, P. G.; Johansson, F.; Lievens, C.; Franz, R. *Packag Technol. Sci.* 1997, 45-58.

[176] Santos, A. S. F.; Manrich, S. In: *Brazilian Polymer Congress.* Associação Brasileira de Polímeros: São Carlos, SP, 2001, 6th.

[177] Devlieghere, F., *et al. Food Addit. Contam.* 1997, 14, 671-683.

[178] Pierce, D. E.; King, D. B.; Sadler, G. D. In: *Plastics, Rubber, and Paper Recycling*; American Chemical Society: New York, 1995, 458-471.

[179] Franz, R.; Welle, F. *Deut. Lebensm-Rundsch.* 1999, 10, 424-427.

[180] Nichols, C. S.; Moore, C. U.S. Patent 5,876,644, 1999.

[181] Erema GmbH. *Recycling News.* Linz, 2000.

[182] Christel, A.; Borer, C.; Hersperger, T. WO 01/34688, 2001.

[183] Rule, M. U.S. Patent 6,103,774, 2000.

[184] Rudiger, F. U.S. Patent 6,436,322, 2002.

[185] Kosior, E. BR Patent P.I. 0,014,094-5, 2002 (in Portuguese).

[186] Hayward, D. W. BR Patent P.I. 0,017,236-7, 2003 (in Portuguese).

[187] Hayward, D. W.; Martin, A. S.; Scholoss, F. M. BR Patent P.I. 9,708,284-8, 2000 (in Portuguese).

[188] Bayer, F. *On the United Resource Recovery Process (URRC)*. São Paulo: Instituto Adolf Lutz, 2000.

[189] *New technology allows recycled PET for food direct contact*. Cempre Informa, 2000, 4 (in Portuguese).

[190] Brownscombe, T. F.; Chuah, H. H.; Diaz, Z.; Fong, H. L.; June, R. L. F.; Kevin, L.; Semple, T. C.; Tompkin, M. R. BR Patent P.I. 9,602,163-2, 1998 (in Portuguese).

[191] *Recycling of soluble plastics by selective extraction*. Fraunhofer Institute Verfahrenstechnik Und Verpackung: Freising, 1999.

[192] Schwartz Jr, J. A. U.S. Patent 6,147,129, 2000.

[193] Louise A. Defresne, L. A.; Agrawaí, R. D.; Moore. F. BR Patent P.I. 9,705,128-4, 1999 (in Portuguese).

[194] Al-Ghatta, R. A. U.S. Patent 5,049,647, 1991.

[195] Schloss, F. M. U.S. Patent 5,824,196, 1998.

[196] Ford, T. *Plastics News*. September1994, 2.

[197] *Scrap Processing and Recycling*. January/February 1995, 28.

[198] Smith, S. S. *Plastics News*. March 1997, 5.

[199] Center for Food Safety & Applied Nutrition *Recycled plastics in food packaging*. 2000, http://www.fda.gov.

[200] Manrich, S.; Santos; A. S. F.; Agnelli, J. A. M. BR Patent P.I 002,993, 2004 (in Portuguese).

[201] Santos, A. S. F. PhD. thesis. Federal University of São Carlos: São Carlos, SP, 2004 (in Portuguese).

[202] Bühler AG. *Popis B2B procesu švýcarské firmy Bühler*. April 2002. http://www.petrecycling.cz/novinkysvet.htm.

[203] Thorsheim, H. R; Armstrong, D. J. *Chemtech*. August 1993, 55.

[204] Manrich, S. *Polym. Recy.* 2000, 5, 213-222.

[205] Manrich, S.; Ravazi, R.; Danella Jr, O. J.; Santana, R. M. C.; Manrich, S. In: *Global Symposium on Recycling, Waste Treatment and Clean Technology*; Madrid, 2004, II, 1671-1680.

[206] Santana, R. M. C.; Manrich, S. In: *International Conference on Polymer Modification, Degradation and Stabilization*; Budapest, 2002, 2nd.

[207] *Celulose & Papel*. 2004, 20, 3 (in Portuguese).

[208] *O papel*. 2004, LXV, 98 (in Portuguese).

[209] Desai, S. C. *Pop Plast. Packag.* 1994, 39, 45-48.

[210] Mannar, S. M. European Patent 605,938 A1 940,713, 1993.

[211] Ohno, A.; Ishige, A.; Koyama, H.; Asami, K. European Patent 685,331 A1 951,206, 1995.

[212] Lin, A. F. U.S. Patent 005,552,011 A, 1996.

[213] Hibiya, T.; Miki, T. European Patent 0,795,399 A1, 1997.

[214] Amon, M. U.S. Patent 66,183,856, 2001.

[215] Williams, R.C. *et al.* U.S. Patent 6,150,005, 2000.

[216] Goettman *et al.* U.S. Patent 6,171,443, 2001.

[217] Squier, J. A. H.; Osgood Jr, R. W.; Perez, K. B.; Williams, D. R. US Patent 6,787,217, 2004.

[218] Nagō, S.; Nakamura, S.; Mizutani, Y. *J. Appl. Polym. Sci*, 1992, 45, 1527-1535.

[219] Ito, K. *et al.* In: *Polymer Processing Symposium*, Guimarães, 2002, 18th.

[220] Danella Jr.; O. J.; Santana, R. M. C.; Manrich, S. *J. Appl. Polym. Sci.* 2003, 88, 2346-2355.

[221] Corrêa, A. C. M.S. thesis. University of São Carlos, São Carlos, SP, 2005 (in Portuguese).

[222] Santi, C. R. M.S. thesis. University of São Carlos, São Carlos, SP, 2005 (in Portuguese).

[223] Manrich, S.; Herrera, J. C.; Rosalini, A. C.; Acconci, C. BR Patent M.U. 7901580, 2005 (in Portuguese).

[224] Santana, R. M. C.; Manrich, S. *Prog Rubber Plast Recy Technol.* 2002, 18, 99-109.

[225] Kaminska, A. Kaczmarec, H. Kowalonek, J. *Eur. Polym. J.* 2002, 38, 1915-1919.

[226] Rangel, E. C.; Gadioli, G. Z.; Cruz, N. C. *Plasm. Polym.* 2004, 9, 35-48.

[227] Manrich, S.; Santana, R. M. C. In: *Meetings on Macromolecules.* Prague, 2005, C67, 23rd.

[228] Nakae, H.; Inui, R.; Hirata, Y.; Saito, H. *Acta Mater.* 1998, 46, 2313-2318.

[229] Sasse, F.; Emig, G. *Chem. Eng. Technol.* 1998, 21, 777-789.

[230] Mancini, S. D. Ph.D. thesis, University of São Carlos, São Carlos, SP, 2002 (in Portuguese).

[231] Karayannidis, C. P.; Chatziavgoustis, A. P.; Achilias, D. S. *Advan. Polym. Technol.* 2002, 21, 250-259.

[232] Goje, A. S.; Chauan, Y. P.; Mishra, S. *Chem. Eng. Technol.* 2004, 27, 790-799.

[233] Klum, U.; Kizan, A. *Polym. Advan. Technol.* 2002, 13, 817-822.

[234] Modesti, M.; Simioni, F. *Polym. Eng. Sci.* 1996, 36, 2173-2178.

[235] Hayashi, F; Omoto, M.; Ozeki, M. Imai, Y. In: *Advances in Plastics recycling series*; Frish, K. C.; Klempner, D.; Prentice, G.; Ed.; Technomic Publishing Company: Lancaster, PA., 1999, 1, 223-240.

[236] Ramsden, M.J.; Phillips, J.A. *J Chem Tech Biotechnol.* 1996, 67, 131-136.

[237] Goto, M.; Koyamoto, H.; Kodama, A.; Hirose, T.; Nagaoka, S.; Mc Coy, B.J. *AIChE Journal.* 2002, 48, 136-144.

[238] Williams, E.A.; Williams, P.T. *J Chem Tech Biotechnol.*1997, 70, 9-20.

[239] Pinto, F.; Costa, P; Gulyurtlu, I; Cabrita, I. *J. Anal. Appl. Pyrol.* 1999, 51, 57-71.

[240] Slapack, M. J. P.; Van Kasteren, J. M. N.; Drinkenburg, B. A. A. H. *Polym. Advan. Technol.* 1999, 10, 596-602.

[241] Buekens, A. G.; Huang, H. *Resour Conserv Recy.* 1998, 23, 163-181.

[242] Masuda, T.; Kuwahara, H.; Mukai. S.R.; Hashimoto, K. *Chem. Eng. Sci.* 1999, 54, 2773-2779.

[243] Kaminsky, W.; Kim, J. S. *J. Anal. Appl. Pyrol.* 1999, 51, 57-71.

[244] Williams, P. T.; Bagri, R. *Int. J. Energy Res.* 2004, 28, 31-44.

[245] Kaminsky, W.; Predel, M.; Sadiki, A. *Polym. Degrad. Stab.* 2004, 85, 1045-1050.

[246] Yoshioka, T.; Grause, G.; Otani, S.; Okuwaki, A. *Polym. Degrad. Stab.* 2006, 91, 1002-1009.

[247] Masuda, T.; Kushino, T.; Matsuda, T.; Mukai, S.R.; Hashimoto, K.; Yoshida, S. *Chem. Eng. J.* 2001, 82, 173-181.

[248] Maraghi, R. In: *Plastics waste management*; Mustafa, N.; Ed.; Marcel Dekker, Inc.: New York, 1993, 369-405.

[249] Snyder, R.H. *Scrap tires: disposal and reuse*; Society of Automotive Engineers, Inc.: Warrendale, PA, 1998, 47-56.

[250] Alexander, J. H. In: *Defense of garbage*; Praeger publishers: Westport, CT, 1993, 147-166.

[251] Abu-Hijlev, B. A-K.; Mousa, M.; Al-Dwairi, R.; Al-Kumoos, M.; Al-Tarazi, S. *Energy Convers Mgmt.* 1998, 39, 1155-1159.

[252] Holmgren, K.; Henning, D. *Resour. Conserv. Recy.* 2004, 43, 51-73.

[253] Huang, C-M.; Yang, W-F.; Ma, H-W.; Song, Y-R. *Waste Manage.* 2006.

[254] Henning, D.; Amiri, S.; Holmgren, K. *Eur. J. Oper. Res.* 2006 *(in press)*.

[255] (2005). www.cgtee.gov.br (in Portuguese).

[256] Holmgren, K.; Henning, D. *Resour. Conserv. Recy.* 2004, 43, 51-73.

[257] Ferrão, P.; Amaral, J. *Technol. Forecast Social Change.* 2006, 73, 277-289.

[258] Correnti, A.; Bocchino, M. Filippi, S.; Magagnini, P.L.; Polacco, G.; La Mantia, F.P. *J. Appl. Polym. Sci.* 2005, 96, 1716-1728.

[259] Saar, S.; Stutz, M.; Thomas, V. M. *Resour Conserv. Recy.* 2004, 41, 15-22.

[260] Massura, A. C.; Souza, E. A. M.; Crochemore, G. B. BR Patent PI 0,202,303-2, 2002 (in Portuguese).

[261] (2005). www.andef.gov.br (in Portuguese*)*.

[262] Lee, C-H., Chang, C-T.; Tsai, S-L. *Resour. Conserv. Recy.* 1998, 24, 121-135.

[263] Machado Neto, J.G. *Pesticide packages disposal.* University of São Carlos: São Carlos, SP, 2004 (in Portuguese).

[264] Kaminsky, W.; Hartmann, F. *Angew Chem. Int. Ed.* 2000, 39, 331-333.

INDEX

G

H